Batteries for Automotive Use

Philip Reasbeck OBE, BSc, PhD, FRIC, CChem, CEng
Formerly Chief Scientist and Director of Research
Lucas Industries plc, Birmingham, England

and

James G. Smith BSc
Principal Research Officer
Lucas Advanced Vehicle Systems Development
Birmingham, England

RESEARCH STUDIES PRESS LTD.
Taunton, Somerset, England

JOHN WILEY & SONS INC.
New York • Chichester • Toronto • Brisbane • Singapore

RESEARCH STUDIES PRESS LTD.
24 Belvedere Road, Taunton, Somerset, England TA1 1HD

Marketing and Distribution:

Australia and New Zealand:
Jacaranda Wiley Ltd.
Sydney Office, Suite 4A, 113 Wicks Road, North Ryde, NSW 2113, Australia
Canada:
JOHN WILEY & SONS CANADA LIMITED
22 Worcester Road, Rexdale, Ontario, Canada

Europe, Africa, Middle East and Japan:
JOHN WILEY & SONS LIMITED
Baffins Lane, Chichester, West Sussex, UK, PO19 1UD

North and South America:
JOHN WILEY & SONS INC.
605 Third Avenue, New York, NY 10158, USA

South East Asia:
JOHN WILEY & SONS (ASIA) PTE. LIMITED
2 Clementi Loop #02-01
Jin Xing Distripark, Singapore 129809

Library of Congress Cataloging-in-Publication Data

Reasbeck, Philip, 1923-
  Batteries for automotive use / Philip Reasbeck and James G. Smith.
    p. cm. -- (Electronic & electrical engineering research
studies. Power sources technology series ; 2)
  Includes bibliographical references and index.
  ISBN 0-471-95742-9 (John Wiley : alk paper). -- ISBN
0-86380-176-5 (Research Studies Press : alk. paper)
  1. Automobiles--Batteries. 2. Lead-acid batteries. I. Smith,
James G., 1940- II. Title. III. Series.
TL272.R39 1997
629.25'42--dc21   97-3760
                             CIP

British Library Cataloguing in Publication Data
A catalogue record for this book is available from the British Library.

ISBN 0 86380 176 5 (Research Studies Press Ltd.) *[Identifies the book for orders except in America.]*
ISBN 0 471 95742 9 (John Wiley & Sons Inc.) *[Identifies the book for orders in USA.]*

Printed in Great Britain by SRP Ltd., Exeter

Editorial Foreword

The aim of the Power Sources Technology Series is to identify areas where significant changes have occurred and to provide an up-to-date picture of the technology in these areas.

One such area is the field of automotive batteries. The old name for the car battery - SLI or Starting, Lighting and Ignition - described accurately the functions of the battery at that time. The modern car is much more complex and the list of electrical equipment now fitted is almost endless and imposes varying loads on the battery. There is even a quiescent load on the battery when the car is not being used from, for example, computer and security equipment. Meanwhile, the design of other electrical installations in the car has not stood still and significant advances have been made in the charging and charge control equipment for the battery.

The modern automotive battery is therefore very different from the traditional 'black box', New materials and designs have resulted in a change not only in external colour but in weight and performance characteristics; for example, the energy density (especially at high discharge rates) is considerably improved and the loss of capacity when the car is left standing has been significantly reduced.

This book follows the changes that have taken place and gives the state-of-the-art of modern automotive batteries. The two authors are well known in the industry. Dr Philip Reasbeck is a former Managing Director of Lucas Batteries and Director of Research for the Lucas Research Centre, and Jim Smith has spent many years working on electrical systems in automotive batteries in the Lucas Advanced Vehicle Systems Development Department. They have written a very readable book full of information and data. Lead-Acid and Nickel-Alkaline technologies are discussed in detail. The final chapter examines future developments both from the point of view of possible further changes in application requirements and in relation to use of the newer battery technologies. The book will be of interest to all battery technologists and to all those engaged in automotive engineering, both in Industry and in Universities.

N. E. Bagshaw
Editor
February 1997

Preface

The major purpose of this book is to introduce automotive electrical engineers and serious students of battery technology to the field of batteries specifically designed for road-vehicle electrical systems. Vehicle electronics and the associated power systems are rapidly becoming more complex and are demanding understanding at a higher level than previously required. The one item that seems to defy simple understanding is the battery. This is evidenced by the difficulty in finding a satisfactory battery model for system simulation or even in finding a totally accurate way of determining the state of charge. It is hoped that by explaining the battery chemistry and construction, at least the excuses for these difficulties will be better understood.

The common-or-garden lead-acid SLI (Starting, Lighting and Ignition) battery is in fact a remarkable example of electrochemical, metallurgical and production engineering and part of the purpose of this book is to illustrate this fact. The solid, unglamorous, plastic container with two terminals hides a wealth of technology and a fascinating array of chemical systems, many of which are still incompletely understood. That such an item can be produced reliably in very large numbers at a competitive price says a great deal about the skill and ingenuity of the scientists and engineers who have been involved with the development of the battery and its related systems.

It is important to note that the lead-acid battery still dominates the vehicle SLI field at a time when many new battery systems are appearing, spurred on by the desirability of electric vehicles and by the wealth of small electronic products that depend on battery power. Though several attempts have been, and are being made, to topple lead-acid batteries from this position (see Chapter 7), none has yet been successful. The lead-acid system has indeed several distinct advantages which suit it for vehicle SLI duties. These can be listed as:

1. The ability to deliver the very large currents required for vehicle starting in a reliable manner.

2. The high cell-voltage (nominally 2.0 V per cell) compared with competing systems. Thus there are fewer cells per battery for a given system voltage.

3. The availability, at relatively low cost, of all the necessary materials. In addition, lead and its alloys are amenable to casting and other familiar processes, thereby permitting cheap fabrication of the requisite parts.

4. The system behaves well with a simple voltage-limited charging system and gives a good service life in the typical vehicle duty.

5. The chemical stability of the components over a convenient range of operating temperatures.

6. The historic availability of the battery at a time when the automotive industry was expanding rapidly. (The large investment in lead-acid SLI manufacturing facilities has doubtless been one important inhibiting factor to any competition.)

For these reasons, the bulk of this book is devoted to explaining the functioning, construction, and manufacture of the lead-acid battery. The authors do so in the firm belief that the lead-acid battery has not only a distinguished past but also has an expanding future role.

The only other system which has been commercially used for road vehicles is the pocket type nickel-cadmium battery, principally in buses and other public service vehicles with large standing loads. A special chapter is devoted to this particular technology. However, the much higher cost of the nickel-cadmium technology has made it easy prey to improved lead-acid batteries.

Automotive batteries are mass-produced like any other vehicle component and the structure of lead-acid batteries has been particularly influenced by production-engineering considerations. The need to recycle most of the battery materials, for both economic and environmental reasons, has had a tremendous impact, and enhanced public sensitivity to health and safety issues in general is having equally important effects on battery production, use, and disposal.

It may be said that the special battery industry which was set up to support the automotive electrical system now has a life of its own and has given birth to a variety of subsidiary activities. Due comment is made in the text on the commercial importance of the battery industry. At the time of writing, the industry is undergoing a series of changes in relationships, as many honoured names disappear and new ones emerge. Besides these organisational changes, there have been over the past three decades very important technical changes, particularly in the use of new materials and new fabrication methods. Low-maintenance and maintenance-free batteries have become commonplace, and steady improvements in life and high-rate performance have taken place.

It is felt opportune that a new book on automotive batteries should be published when there is a renewed interest in automotive systems and their functional architecture. There are already classical works in their own right by Vinal, Bode, and Dasoyan and Aguf, not to mention the comprehensive work by Barak. However, it is hoped that this book will satisfy a special need for those approaching the topic from a pragmatic technical viewpoint, and at the same time will help those wishing to enquire more deeply to ask intelligent questions, and to find useful answers.

Acknowledgments

Acknowledgments are particularly due to Mr J. M. Ponsford, formerly Manufacturing Manager of Lucas-Yuasa Batteries, and now retired, for his very considerable contribution to the chapters concerned with matters of Manufacturing, Design, and Basic Principles.

Secondly, we wish to thank Mr R. E. Blakeley for producing many of the diagrams.

We are also indebted to Dr D. N. Wilson, Director of The Lead Development Association, for permission to use the LDA archives for material for Chapter 4.

Thanks are due to the following companies for providing photographs or illustrations, and for permission to use such items in this book :

Lucas Yuasa Batteries Ltd, Birmingham, UK

Alcad Ltd, Middlesex, UK

Britannia Refined Metals Ltd, Gravesend UK

Cominco Ltd, Ontario, Canada

Elbak Maschinenbau G.m.b.h., Graz, Austria

GNB Inc. Milwaukee, Minnesota, USA

MAC Engineering and Equipment Co. Ltd, Michigan, USA

T.B.S. Engineering, Cheltenham, UK

Wirtz Manufacturing Co. Inc., Michigan, USA

We are specially indebted to colleagues in various Divisions of Lucas Industries plc., particularly from Lucas-Yuasa Batteries and the former Advanced Engineering Centre for their discussions, support and encouragement over many years, and we would like to thank the Proving Laboratory for information on battery testing and vehicle performance.

Our thanks are due to Mrs V. Wallace, Managing Director of Research Studies Press Ltd., for her very considerable help and guidance, and particularly for her patience over several years, without which this book would never have been finished.

Norman Bagshaw, Editor of this book series, gave us considerable help with technical matters, for which we thank him.

Finally, we wish to thank Mr Bob Woolley, of Frederick Woolley Ltd, Birmingham, for his kind permission to use the word processing facilities of the Frederick Woolley Learning Centre.

P. Reasbeck

J. G. Smith

February 1997

Contents

CHAPTER 1

Historical Background to the Development of Vehicle Electrical Systems and Secondary Batteries

The electrical system of a modern road vehicle is a vital part of the functional unit. Its performance is critical to mobility, safety and comfort, and any loss of electrical power may well render the vehicle immobile and useless. Since the system is required to be active before the engine-driven generator can function, some kind of storage battery has to be fitted.

As many motorists are aware, the performance of the lead-acid battery which usually supports the starting, lighting and ignition system often determines whether or not a vehicle can be used. However, the appearance and location of the battery often belie its importance. An inscrutable black box with two terminals, it can sometimes only be reached by heroically agile service mechanics.

We shall consider in this opening chapter how the battery is related to the vehicle system, the loads it has to support and the arrangements for replacing the energy it gives up on discharge. We shall consider especially the lead-acid battery in this context. Though it does not have the field entirely to itself, its particular characteristics and its availability in a cheap and convenient form at a time when the vehicle industry was booming have given it an extremely strong position. Very often, the only secondary battery of which the average person is aware is that which starts his car.

In this introductory chapter, we shall first look at the system which the battery is required to power and then at the battery itself.

1.1 Road Vehicle Electrical Systems
The pre-1900 motorist was in no need of electrical complications. His ignition source was a flame-heated tube, and oil lamps gave sufficient illumination for the speeds at which he travelled, should he be bold enough to drive at night. There were few other vehicles to avoid and the noise he made gave ample warning of his approach.

Nevertheless, many of the basic electrical components of the systems we know today were already available. Coil ignition, the electric light bulb and electric motors were commercially available and being used for other purposes. Various primary batteries and the lead-acid secondary battery were being used as sources of power. Presumably, familiarity with things mechanical rather than with things electrical made the early car owner suspicious of using electric power as an aid.

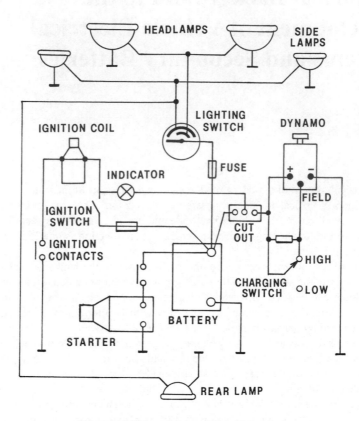

Fig.1.1. 1920s vehicle wiring diagram

Though stationary gas engines used electric ignition in the 1860s, the first use of electricity on road vehicles was for lighting kits powered by primary batteries, or occasionally secondary lead-acid batteries (Hawkins, 1.1). Oil lamps continued to be used until the 1920s, but simple lighting outfits consisting of two lamps, a switch, cables and battery were the embryos of the complex systems of today. Almost any of the primary batteries in use could support the moderate currents required and Leclanché 'dry' batteries were popular.

The advent of electric ignition, in the form of an engine-driven magneto, did little to change the situation, since the high tension supply was produced directly by mechanical means. The real change came in 1910 when Cadillac in America announced 'the car without a crank' and the electric starter made its debut

(Schallenberg, 1.2), though crank handles were thoughtfully provided for another five decades. This original example used a 6 volt lighting system but had four batteries in series to work the 24 volt starter (a system remarkably similar to that discussed for modern applications in a later chapter).

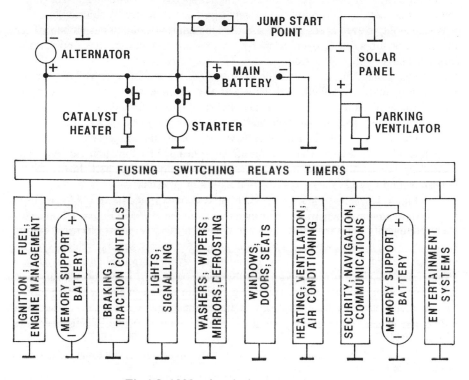

Fig.1.2. 1990s electrical system schematic

Starting power requirements were two orders of magnitude higher than those for lighting applications and it became obvious that a special type of battery was necessary for road vehicle duties. Moreover, since the starting power had to be available on demand, the reliability of a secondary battery kept at full charge by an engine-driven dynamo was clearly appreciated.

Figure 1.1 shows a car wiring diagram of the 1920s, which though relatively simple, had most of the functions required today. Over the next four decades, luxuries such as windscreen wipers and washers, screen heaters, interior heaters and ventilators together with external safety indicators and lighting fixtures became standard fittings as traffic conditions changed and the use of vehicles in more adverse weather conditions became accepted.

The next major change occurred in the 1960s with the advent of semiconductor devices which could operate at the temperatures found in engine compartments (up to 125°C). In particular, small, cheap rectifier diodes with high current capacity paved the way for the replacement of the DC dynamo by the more efficient AC alternator. The automotive version of the alternator has specific outputs in excess of 120 W/kg compared with about 45 W/kg for the dynamo. In addition, the alternator characteristics enable it to run at higher speeds without damage. The pulley ratios on the belt drive from the engine crankshaft can therefore be adjusted to give a reasonable output even at engine idle speeds. This improved output has probably been the salvation of systems overloaded by various accessories and has considerably improved battery life expectation.

Semiconductors have also been responsible for much better voltage control, new ignition and engine fuelling systems and also a whole new spectrum of controls and instrumentation to meet the demands of current emissions and safety regulations. Communications, navigation and diagnostics systems are now being added to the list.

Legislative trends in particular have had profound influences on electrical system and battery design. The need for effective headlamps, direction indicators and the disappearance of parking light requirements all played their part. In the 1990s, emission controls and catalyst systems will have similar effects.

Figure 1.2. shows the kind of electrical system expected in the 1990s. It has to be shown in a more schematic manner than that from the 1920s in Figure 1.1 since the sheer complexity does not allow the same detail on a single page. Electrical control units involving microprocessors are now common features in several vehicle systems and these make their own specific demands in terms of reliable battery support (sometimes achieved by using primary batteries once again). The battery also has an important function in acting as a smoothing capacitor and as an energy sink to protect the electronic systems from excessive voltage transients, and for this reason it is extremely ill advised to run a modern vehicle engine if the battery becomes disconnected.

A look into the more advanced aspects of vehicle electrical systems is attempted in Chapter 7.

1.2 The Basic Requirements of Vehicle Electrical Systems

The road vehicle's electrical system consists of a generator, a storage battery, voltage control and protective devices and the electrical loads. The latter items are the system's raison d'être and they will be examined first.

Table 1.1 shows the loads that can be expected in modern passenger vehicles. The list in Table 1.1 is far from exhaustive. The currents listed are taken from a 'nominal' 12 volt system, whose working voltage may actually vary from about 8 (on starting) to 14.5 (with a fully charged battery and light loads). Commercial vehicles, public service vehicles (buses and coaches) and certain custom vehicles such as ambulances and police cars have specialised loads which make much heavier demands, sometimes requiring the use of two batteries, an extra generator or the use of a nominally 24 volt system. The loads on passenger car systems have increased over the recent decades as shown in Figure 1.3 (Holt and Williams, 1.3).

Table 1.1. Typical loads on a passenger vehicle system

Load	Power (W)	Current (A)
Starter Motor	1000-5000	200-1000
Headlamps	100-200	10-20
Side/Trail Lamps	50	4
Direction Indicators	50	4
Instruments	20-30	2
Fog Warning Lamps	200	15
Windscreen Wipers	100	8
Windscreen Washers	20	2
Window Operation	100-200	10-30
Seat Adjustment	100-200	10-30
Heater Blower	20-100	2-10
Air Conditioning	500-1000	40-80
Horns	200	15
Radio/Stereo	20-100	2-10
Rear Screen Heater	180	15
Front Screen Heater	500	40
Engine Controls	12-60	1-5
Fuel Injection	200	15

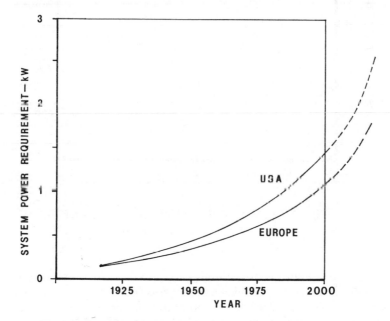

Fig.1.3. Growth of loads on passenger car electrical systems

6

Some of the loads, such as wiper motors or heating elements, are tolerant of wide voltage variations, while in other cases the voltage variations must be limited for proper operation. A quite small change can have a profound effect on the light output from tungsten filament lamp bulbs, an effect commonly noticed when increasing engine revolutions (and hence alternator output) with several heavy loads switched on. The effect of operating voltage on bulb output and life can be seen in Figure 1.4.

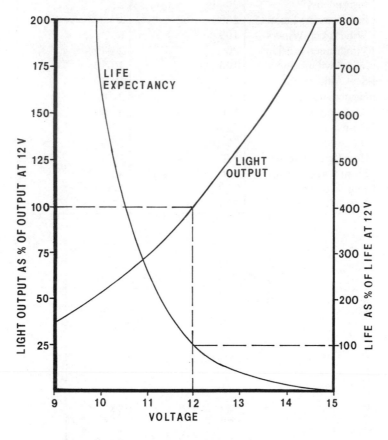

Fig.1.4. Effect of voltage on bulb life and brightness
(12 V rated headlamp bulb)

It can be seen that a change from 12.5 V to 13 V increases lamp output by 13% but reduces life by 35%. There are similar effects of voltage variation on heaters, indicators, etc., but the results are not so critical in the usual range encountered. Ignition coils can be designed to work at quite low voltages, and indeed many ignition systems use a low voltage coil with a series ballast resistance, the latter being shorted out when the starting load reduces the overall system voltage.

Some types of dashboard instrument (e.g. fuel gauge indicators) were very sensitive to voltage variation and a crude 'energy regulator' relying on heating of a bi-metal strip which periodically broke the circuit with a mark/space ratio dependent on the applied voltage. Current trends with modern instrumentation rely on internal voltage regulation and voltage transient protection which accommodates all but the most catastrophic of voltage excursions.

As will become clear elsewhere, the battery itself acts as a voltage smoothing device, its current/voltage characteristic approximating to that of a capacitor of several farads equivalent capacity. The system designers rely on this characteristic to assist other smoothing devices, and for this reason, a modern vehicle should never be run with battery disconnected. (The actual disconnection of a battery while running may also result in serious damage by causing alternator load dumping transients fatal to semiconductor devices.)

1.3 Battery Charging Systems

In virtually all automotive systems the battery and the alternator or dynamo are placed in parallel across the system and the battery is kept at the alternator output voltage which is also applied directly to the loads. Proper voltage regulation is not only necessary for the protection of system loads, but also for the correct functioning and long life of the battery. At an early stage, it was found that too high a charging voltage resulted in battery overheating, heavy water loss and drastically shortened life. Too low a charge voltage gave incomplete recharging with resulting continual loss of battery capacity.

The actual voltage limit for the best output and life optimisation is discussed further in Chapter 3. It will be sufficient to say that a nominally 12 volt lead-acid battery will be placed across a system limited to a maximum of around 14.5 volts, and that limit will only be reached when the battery is nearly fully charged and the generator output is not being consumed by the other system loads.

The original Lucas G80 dynamo, fitted in 1912, had a clutch to disengage its drive when the engine speed threatened to drive the output too high (Nockolds, 1.4). Other machines used a belt drive with the tension so adjusted as to slip when the speed or output got too high. A more developed attempt at voltage regulation used the 'three-brush dynamo' (Judge, 1.5). This was a DC dynamo with a field coil excited by means of current from a third brush on the commutator. As the dynamo current increased, distortion of the main field caused by the armature current reduced the output voltage at the third brush, thus reducing the field current and lowering the output voltage. The position of the third brush could be altered between 'winter' and 'summer' positions, the former giving generally higher output to cope with increased winter loads and the poorer charge acceptance of the battery at low temperature. Sometimes, the same effect was achieved by two values of series resistance selected from a dashboard switch. This arrangement is shown in Figure 1.5a.

A later and more effective scheme was to use a regulator relay controlling

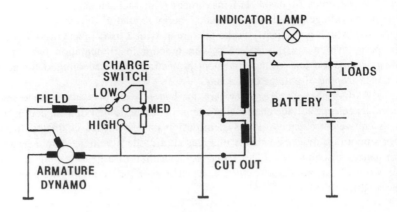

Fig.1.5a. Third brush dynamo charging circuit

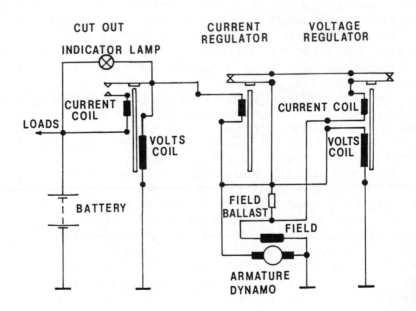

Fig.1.5b. Three relay (current regulator) charging circuit

current through the field coils. This had a high resistance voltage regulating winding which opened the dynamo field circuit when the system voltage reached the desired value. The loss of field then reduced the voltage until the field relay closed once again, the cycle repeating itself at frequencies of up to 100 Hz, the actual values depending on the load and dynamo input conditions. The pulsations of field current and output were smoothed to a great extent by the inductance of the components involved and the capacitive effect of the battery. Later versions of field relays also incorporated a winding in series with the dynamo output. This assisted the shunt (voltage) winding in opening the field contacts when the output current was high (e.g. when recharging an exhausted battery), and thereby reduced the output voltage, preventing overloading of the dynamo and the system in general (Young and Griffith, 1.6).

Some means of preventing battery discharge through an inactive dynamo had to be provided. In the days before cheap rectifiers of adequate current capacity, this was achieved using a 'cut-out' relay of similar construction to the regulator relay and with both series and shunt windings. The armature had contacts which, when open, isolated the dynamo from the rest of the system. On starting, the dynamo voltage would increase until the current through the high resistance shunt winding was high enough to close the contacts. The voltage at which this occurred was chosen to be safely in excess of that to be expected from the battery. Should the dynamo voltage fall below the system voltage held by the battery, the reverse current through the series winding acted in opposition to the shunt winding and assisted the rapid opening of the contacts.

The actual voltages at which the various regulator and cut-out events took place were adjusted mechanically by the spring tensions on the relay armatures. Sometimes a bi-metal spring was provided for the regulator armature so that the system voltage could be raised at lower temperatures, thus compensating for higher on-charge voltages when cold and also for reduced charging efficiency. Usually, both regulator and cut-out were placed in a common enclosure together with the system fuses. The general arrangement of the circuitry is shown in Figure 1.5b. An indication of the system effectiveness can be seen in Figure 1.6.

The final version of the electromechanical voltage controller, as typified by the Lucas RB340, introduced in the early 1960s, used a third relay in addition to the voltage regulator and cut-out. This was responsible for restricting the current output and giving extra voltage protection.

Electromechanical systems of the type described had several weak points. Wear of the relay contacts and spring tension variations caused drifts in the voltage settings and limited the effective life. Later versions needed complex and careful adjustment. The sparking contacts were also good sources of electrical interference. However, once set up, they coped reasonably well with the demands of the time and serious failure was not very frequent.

The semiconductor revolution of the 1950s and 1960s first made its presence felt in the automotive field by the changes made possible in current generation and voltage regulation. Heavy current rectifier diodes of small size and adequate reliability at a reasonable price became available and these enabled the AC alternator to supplant

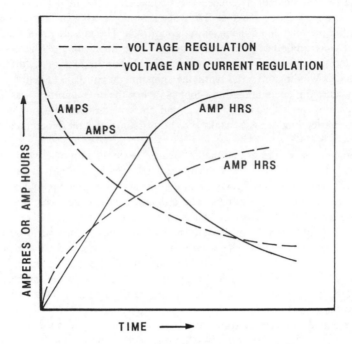

Fig.1.6. Dynamo charging characteristics

the DC dynamo. Alternators were already in limited use on motorcycles where a small battery and load could be supplied economically by a metal oxide rectifier. (Heavy loads such as headlamps were fed directly with raw AC current.) Large public service vehicles (buses) also used alternators to supply their exceptional loads but they had ample room to accommodate the bulky metal oxide rectifiers used for larger currents. However, the small silicon diode which would survive in the service environment (particularly at the under bonnet temperatures encountered) made the alternator available to the whole range of road vehicles. The advantages of the alternator are illustrated in Figure 1.7.

The output characteristics of the alternator make for considerably simpler control circuitry. Its internal inductance limits the maximum output current available and so current limiting devices are not necessary, at least for the protection of the machine itself. A bi-stable switching circuit, triggered by a voltage sensing Zener diode, is all that is needed to control the field current and hence the output voltage. Such a circuit, now realised in integrated thick film form, has no moving parts and is more stable and longer lived than the old mechanical regulators. The output and control circuitry for an alternator is shown in Figure 1.8.

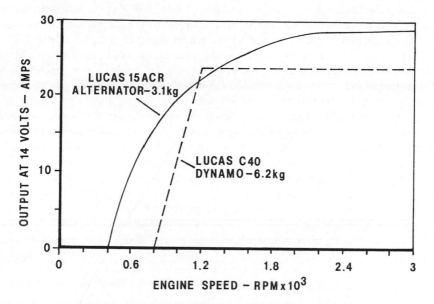

Fig.1.7. Comparison of alternator and dynamo

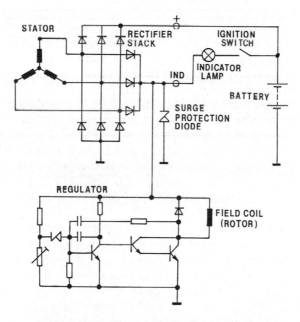

Fig.1.8. Alternator with electronic regulator

The beneficial effect of the alternator has manifested itself in the improved life of most vehicle batteries. An improvement of battery life over the last three decades has not been due merely to changes within the battery itself but also to the better voltage regulation and better charge current availability even in engine idle situations.

In virtually all vehicle systems, the chassis and bodywork are used as the current return to one pole of the battery (i.e. as one rail of the power supply) and an interesting aspect of the vehicle electrical systems, which has been the source of some controversy in the past, is exactly which side of the battery should be connected to earth, i.e. to the metal of the vehicle chassis or bodywork. Very early systems used a negative earth, but it was found that the corrosion of certain components, particularly battery terminals, could be diminished by using a positive earth. Also the conventional practice of ignition coil connections with a positive earth meant that the central (live) pole of the spark plug became negative and was a more reliable spark emitter when hot. However, since the 1960s, a reversion to negative earth has taken place, better wiring and connector protection having reduced the risk of corrosion, and the ignition advantages have been achieved by other means. (The negative earth was also more compatible with American vehicles which apparently stayed negative earth from the outset.)

1.4 The Development of the Secondary Battery

Volta's discovery of the galvanic cell in 1800 gave the world its first real source of steady and controllable electric current. Other workers soon took up investigation of the chemical effects of the current, and in 1802 Gautherot discovered a remanent voltage on platinum wires which had been used to pass current through a salt solution. Ritter in 1803 repeated the experiment with a variety of metal electrodes and recognised that the surface oxidation of the metals was the source of the voltage remaining when the primary current source was removed. He went on to construct what is now recognised as a 'bipolar' battery of copper discs, separated by absorbent layers soaked in electrolyte (Schallenberg, 1.2).

Grove in 1839 made a secondary cell using platinum electrodes in contact with electrolyte and with hydrogen (negative) or oxygen (positive). This was really a form of fuel cell, as referred to in Chapter 7. Sinsteden in 1854 produced secondary voltages from cells with silver, lead and nickel electrodes in alkaline electrolyte saturated with zinc oxide. He may therefore be credited with embryo silver-zinc and nickel-zinc batteries, and he was known to have experimented with lead in sulphuric acid. Other workers (Sinsteden, 1.7) had already used lead dioxide as positive material in primary cells.

In spite of these early uses of lead electrodes, the practical realisation of the lead-acid battery as a usable secondary system is generally credited to Gaston Planté, an electrochemist employed by the Parisian telegraph manufacturers Christoff et Cie. In March 1860, Planté presented a battery of ten cells to the Académie Française. Each cell consisted of two concentric spirals of lead sheet separated by porous cloth and immersed in dilute sulphuric acid within cylindrical glass containers as shown in Figure 1.9. The active area of the lead sheets was enhanced by a process of charging with current from a Bunsen primary battery and allowing the cells to self-discharge over a long period. This tedious latter stage was soon replaced by forced discharge to save time.

Planté's cells were valued as much for their 2 volt output and their high current discharge capability as for their rechargeability. On discharge, they outperformed all the primary cells then in use and Kirchhoff (1861) suggested their use to smooth the output of dynamos, thus partly anticipating the automotive systems of today. In terms of capacity per unit weight and volume, however, the Planté cells did not match the available primary cells and the formation process of repeated charge and discharge was time-consuming and expensive. Fauré in 1880 found that higher capacities could be obtained by the oxidation and reduction of pastes of lead compounds rather than of the metal itself as in Planté's cells. He produced a construction using pastes of red lead (triplumbic tetroxide) coated onto lead sheets which were then rolled together in the manner of Planté. During the rolling operation, paste often fell off or was squeezed out.

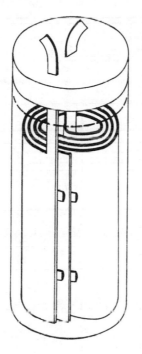

Fig.1.9. Planté battery

The names of Volkmaar, Swan and Sellon are associated with the idea of using perforated sheets of metal and, later, cast grids of lead alloy which held the material in place (Wade, 1.8). A change was also made from Planté's 'Swiss roll' construction to the more familiar flat interleaved structure which made large cells easier to produce.

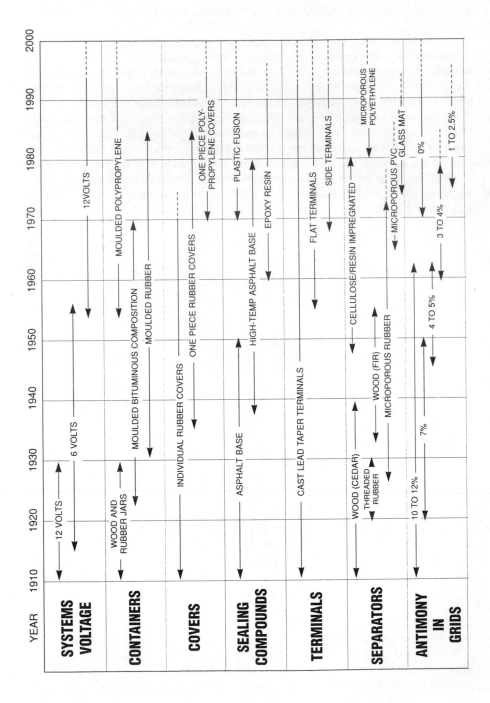

15

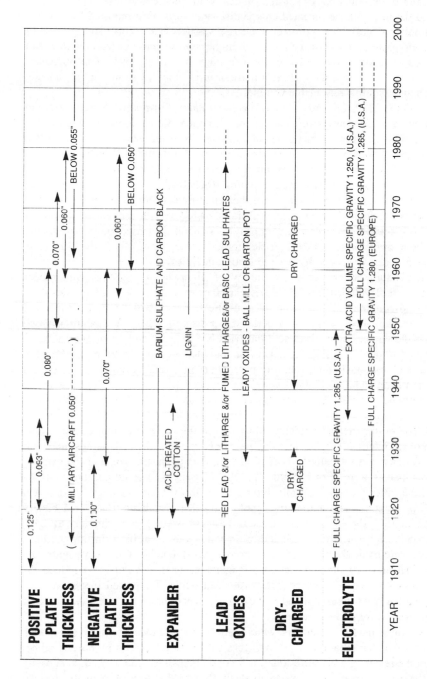

Fig. 1.10. History of battery manufacturing

Fauré's introduction of pasted plates with inherently high surface area eliminated the need for the repeated charge/discharge formation required by Planté's cells and only one prolonged 'formation' charge was necessary. This had the effect of oxidising the paste on the positive plate to a highly porous mass of lead dioxide, and that on the negative was reduced to a spongy form of lead. The increased electrochemical activity of the materials in these forms enabled a substantial capacity to be contained within a reasonable volume.

From the 1880s onwards, lead-acid cells of Fauré's or related designs began to be used in a wide variety of applications ranging from domestic lighting to electrically-propelled road vehicles. (Planté's system was, and still is, used in certain stationary applications where the long life is critical.) A scheme of 'Electric Catering' was even proposed whereby electrical energy, neatly stored in secondary cells, would be delivered to the doorstep each morning (Schallenberg, 1.2). The increased demands, however, taxed the resources of the scientific instrument suppliers who had hitherto made batteries of all types, and specialist battery concerns began to appear.

The Electric Power Storage Company of London, Gottfried Hagen (originally lead smelters) of Cologne and the Electric Storage Battery Company in the USA were among the pioneers of special battery production facilities.

At the turn of the century, a variety of 'lightweight' batteries with rubber containers were developed for the electric cars then much in vogue. Electrodes were made of lead-antimony alloy of improved stiffness, pasted with oxide mixtures and held apart by thin sheets of porous wood. T. A. Edison in the USA was not satisfied with what the contemporary lead-acid batteries had to offer and spent several years and half his fortune in developing nickel-iron and nickel-cadmium batteries, the latter being discovered almost simultaneously by Jungner in Sweden (Jungner, 1.9, Edison 1.10). The nickel-cadmium battery was later destined for vehicle system use where exceptionally heavy loads were encountered, and longer life justified the much higher cost.

Close on the electric vehicle enthusiasm of the early twentieth century came the improved internal combustion engine vehicles with their dynamos and self starters. The demand for batteries increased by orders of magnitude, and the existing battery companies were joined by others such as Lucas in England and Delco in the USA. These latter were specifically interested in the vehicle starting, lighting and ignition applications. The automotive application then dominated lead-acid battery development for three or four decades. Plates were made thinner to give proportionally higher active surface and lower voltage drops at high starting currents. Separators were made from thin slices of porous wood to achieve the same ends, and new terminal and cell interconnection arrangements were developed to reduce the resistive losses in the conductors. The basic lead oxides for pasted plates were no longer supplied from the traditional paint and pigment sources but processes specific to the application were found necessary.

As more modern materials were tried, some unexpected discoveries were made. Early wooden separators of Port Orford Cedar were replaced by porous rubber sheet, but the resulting cells were found to have shorter lives, due to gradual capacity loss by the negative plate. Appropriate analysis showed that lignosulphonates formed by the action of sulphuric acid on the wood had a beneficial action on the negative plate functions. This so called 'expander' action could be produced artificially by adding lignosulphonate derivatives to the negative plate active material during the initial oxide mixing process. Such 'expanders' are now found essential for good low temperature performance and long life, though their exact mode of operation is still not fully resolved.

Rapid expansion of the motor industry brought demands for lighter and cheaper batteries, and battery design has been much influenced by the mass production techniques and supply logistics developed within the automotive industry, as will become clear from later chapters. 'Dry charged batteries', i.e. batteries, nominally in a fully charged state, but devoid of electrolyte during storage, became available. These were safe to transport and store and could be rapidly activated at the point of sale. Maintenance-free designs requiring no topping up during their expected lifetime became generally available and gave more freedom for battery location within the vehicle. Figure 1.10 shows the general chronology of the developments referred to above. The significance of the various items will become clear in later chapters. Future trends are the subject of Chapter 7.

So far, we have concentrated on the lead-acid battery, and most people would recognise no other battery in the automotive application. Nickel-cadmium is still occasionally used where it is believed to be cost effective, particularly in large public service vehicles. The only version used regularly on vehicles is the so-called 'pocket plate' construction developed by Edison. Active materials are contained in nickel-plated steel perforated pockets, linked together to form a plate. Cell and battery constructions are substantially different from the lead-acid versions and are described in detail in Chapter 6. Though the discharge power capability far exceeds that of normal lead-acid cells and there are also significant life advantages, the high cost (six to eight times that of lead-acid) and perhaps also the different charging requirements have so far prevented any more general application in road vehicles.

REFERENCES

1.1 Hawkins Electrical Guide, Volume 9-Sections 2733-2814, Published by T Audel and Co., New York, 1917.

1.2 'Bottled Energy', R H Schallenberg, Published by The American Philosophical Society, 1982 [an excellent account of early battery research and commercialisation].

1.3 'The Future of Vehicle Electrical Power Systems and Their Impact on System Design' by M J Holt and G A Williams, SAE Paper SAE911653, Future Transportation and Technology Conference, Portland, Oregon, 1991.

1.4 'Lucas-The First 100 Years' by H Nockolds, Pub. David and Charles, 1976.

1.5 'Automobile Electrical Maintenance' by A W Judge (5th Edition), Pub. Pitman, 1965.

1.6 'Automobile Electrical Equipment' by A P Young and L Griffiths, Pub. Illife Books, 7th Edition, 1962.

1.7 N J Sinsteden, Annalen Physikalische Chemie, Vol 92, p21, 1854.

1.8 'Secondary Batteries' by E J Wade, Electrician Printing and Publishing Co., 1902.

1.9 W Jungner, Swedish Patent 10,177 (1899).

1.10 T A Edison, German Patent 157,290 (1901) and US Patent 678,722 (1901).

CHAPTER 2

Basic Principles
of the Lead-Acid Battery

2.1 Galvanic Cells and Batteries

Battery technology involves a varying mixture of chemistry, materials science, and mechanical and electrical engineering. It would be too tedious to derive all the required material of this chapter from first principles but it is hoped that readers with an elementary background in these topics will be able to follow the descriptions and arguments involved. In order to help those unfamiliar with the topic, references are made to standard works which provide more detailed descriptions and rigorous derivation of the formulae involved. At the same time, appendices (Appendix 2 and Appendix 3) are provided for those already familiar and wishing to find the appropriate numerical data.

Like living organisms, batteries are made up of individual cells. All cells in a battery are normally identical and each one contributes a characteristic voltage determined by the chemical reactions which take place within the cell. Cell capacity is determined by the weight of chemical reagents which are contained within the cell. The cell units are connected in series or, more rarely, in parallel, to obtain the desired voltage and capacity.

A power producing cell ('galvanic' cell) consists of two electronically-conducting elements or 'electrodes' in contact with an ionically conducting phase, the 'electrolyte'. On the surface of the electrodes in contact with the electrolyte, there occur reactions ('charge-transfer' reactions) whereby electrons are exchanged between the electrodes and ions in the electrolyte. These charge-transfer reactions are the source of the cell's power and define its major characteristics. Charge flows between the electrodes by means of ions in the electrolyte, and the circuit is completed by electron flow between the electrodes via the external circuit. The generalised situation is shown in Figure 2.1 and the actual mechanism of the lead-acid cell will be used as a model for more detailed explanation.

Normally, the electrodes in an electrolytic cell are designated as cathode (negative) and anode (positive). However, in a secondary galvanic cell (i.e. a cell where the active materials are regenerated electrically), the cathode, which is the site

20

of a reduction reaction, is the positive cell terminal and the anode is the site of an oxidation reaction and forms the negative terminal. When a secondary cell is being recharged from an external power source, the positive electrode becomes an anode (as the discharge reactions are reversed) and the negative electrode becomes a cathode. To avoid confusion when talking about secondary batteries, it is best to refer to 'positive electrode' and 'negative electrode', or in battery parlance, 'positive plate' and 'negative plate'.

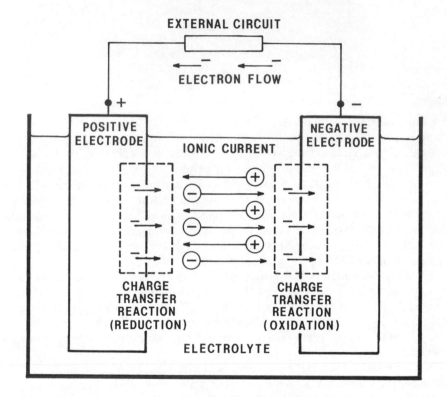

Fig.2.1. Galvanic cell schematic

2.2 Thermodynamic Relations in Galvanic Cells

The study of electrochemical reactions is helped considerably by the useful relation between reaction parameters and easily measurable quantities. Thus the potential of a cell is simply related to the Free Energy (Gibbs Free Energy Function) change in the cell reactions. (The basic thermodynamic functions such as enthalpy, entropy, and Gibbs function or Gibbs free energy are well discussed and related to measurable quantities in standard texts on physical chemistry, e.g., Atkins (2.1), Borrow (2.2)).

Standard Potential, $E_O = -\Delta G_O/n.F$ (2.1)

ΔG_O is the Free Energy change (joules/mole) involved in the cell reaction,
n is the number of electrons transferred per molecule of reaction,
F is the faraday (96,485 coulombs/mole).

The voltage measured between two electrodes in a practical cell is equal to the difference between the electrode/electrolyte potentials at the two electrodes.

Cell Potential, $E_O = E_{+ve} - E_{-ve}$ (2.2)

Since it is impossible to measure the potential at a single electrode/electrolyte interface (another electrode must be introduced to 'make contact' with the electrolyte), standard electrode potentials are, by convention, assumed to be measured in a cell using as its second electrode a standard hydrogen electrode (abbreviated to SHE; an electrode with hydrogen at one atmosphere in contact with an electrolyte having unit-activity hydrogen ions), even though such an electrode is difficult to realise in practice. On a practical level, a cadmium rod immersed in the electrolyte was often used as a rough and ready reference electrode for lead-acid batteries. A more reliable reference electrode for use in sulphuric acid is the mercury/mercurous sulphate electrode (Ives and Janz, 2.3). Figure 2.2 shows the various electrode potentials versus standard and practical reference electrodes in a lead-acid cell.

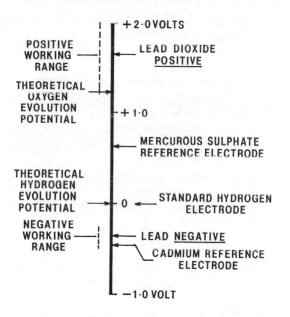

Fig.2.2. Potential relationships in the lead-acid cell

Electrode potentials are related to the Free Energy change of individual electrode reactions (by equation 2.1) in the same way as cell potentials are related to overall cell reactions. Such potentials are also related to the concentration of the ions in solution by the Nernst Equation (Atkins, 2.1). Thus for a reduction reaction

$$O + ne \rightarrow R$$

the potential concentration relationship is

$$E = E_O + RT/n.F.\ln(a_O/a_R) \tag{2.3}$$

where a_O is the concentration (or more precisely the activity) of the oxidised species in the electrode reaction, and a_R is that of the reduced species.

Cell potentials are also temperature-dependent and it may be deduced from equation 2.1 that the cell potential is related to the entropy change in the cell reaction as in equation 2.3

$$\partial E/\partial T = \Delta S/n.F \tag{2.4}$$

where $\partial E/\partial T$ is the cell temperature coefficient (at constant pressure),
and ΔS is the standard entropy change for the electrode or cell reactions concerned.

A similar relation relates the enthalpy change ΔH (joules/mole) with the cell potential and temperature coefficient via the Gibbs-Helmholtz equation:

$$\partial E/\partial T = 1/T(\Delta H/n.F + E) \tag{2.5}$$

Note that all the relationships above hold for a cell working in ideal thermodynamically reversible conditions. When current flows into or out of an electrode, a net reaction takes place which displaces the equilibrium at the electrode surface. The reaction flux is related to the electrode current by Faraday's law.

$$i = n.F.j/A \tag{2.6}$$

where i is the current density at the electrode surface,
j is the reaction flux in moles/s,
A is the electrode surface area.

As is seen later, the current density rather than the absolute current is the parameter which determines the electrode and cell behaviour.

2.3 Cells with Current Flowing

When current flows through a cell, the voltage is displaced from the open-circuit equilibrium value and voltage changes occur which are known as 'polarisation' effects. The effects on voltage are illustrated in Figure 2.3. Note that cell voltage falls below the open-circuit value on discharge and rises above the open-circuit value on charge. These voltage changes diminish the efficiency of a working cell, even if all the charge input can be recovered. They are due to effects classically referred to as 'polarisation' (or occasionally as 'overvoltage') and arise from three main causes:

Activation Polarisation - representing the energy required to overcome the reaction's activation energy barriers This may be subdivided further to account for various rate-limiting processes such as electrocrystallisation (the nucleation and growth of crystalline reaction products).

Concentration Polarisation - representing the changes in electrode potentials brought about by changes in local concentrations as the reaction proceeds.

Ohmic, or Resistance Polarisation - representing the resistive losses in the current path through the cell components.

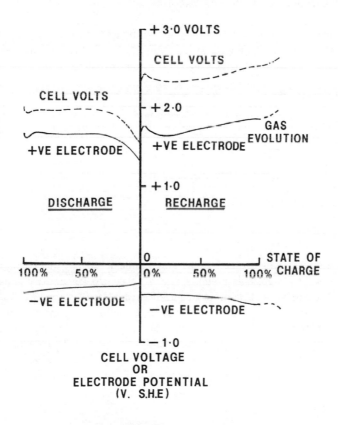

Fig.2.3. Cell voltage and polarisation effects for a lead-acid cell

2.3.1 Activation Polarisation

The theory of reaction kinetics postulates energy barriers to the progress of reactions which can be overcome by thermal energy, or in the case of electrode reactions, by electrical energy input. The energy required is derived from a change in electrode potential from the equilibrium value by an amount, η_a, the activation polarisation. The electrode surface is assumed to be in a dynamic equilibrium with a so-called 'Exchange Current' flowing through the surface in both directions. At the Equilibrium Potential, the current flowing in both directions is equal and no net current flows. It can be shown (Albery, 2.4; Bard and Faulkner, 2.5; Vetter, 2.6) that the effect is related to the inherent speed of the electrode reaction via the following relations:

$$i = i_0 \left(\exp{^{\alpha.n.F.\eta}/_{R.T}} - \exp{^{(1-\alpha).n.F.\eta}/_{R.T}} \right) \tag{2.7}$$

where i_0 is the current density of the 'exchange current' which is flowing at the electrode surface,
α is a constant relating to the shape of the activation energy barriers,
and the other symbols have the meanings ascribed previously.

The nature of equation 2.5, known as the Butler-Volmer equation, allows two approximations which are borne out in experience. At very low current densities and small values of η (below 5 mV), there is a linear approximation

$$\eta = i.R.T/_{i_0.n.F} \tag{2.8}$$

At higher currents and values of η above 200 mV, there is a logarithmic approximation,

$$\eta = {^{R.T. \ln i}/_{\alpha.n.F}} - {^{R.T. \ln i_0}/_{\alpha.n.F}} \tag{2.9}$$

This can be simplified further to

$$\eta = a + b.\ln i \tag{2.10}$$

Equation 2.10 was evolved also by experiment and is usually known as the 'Tafel Equation'. The constants a and b can be related to i_0 and α by equation 2.7. They are commonly measured as the intercept and slope of logarithmic current/voltage plots in cells where activation polarisation is the predominant effect. The exchange current density is sometimes envisaged as the current flowing in and out of the electrode surface in dynamic equilibrium. It has a characteristic value determined by the reaction rate constant and the concentrations at the electrode surface.

$$i_0 = k_s.a_r^{\alpha}.a_o^{(1-\alpha)} \tag{2.11}$$

where k_s is the standard rate constant for the electrode reaction.

It will be appreciated that the higher the value of the reaction rate, the higher the value of the exchange current, and, as can be deduced from equations 2.7 to 2.9, the lower will be the value of activation polarisation, η_a, for a given current density. Thus any influence such as catalysis which lowers activation energy barriers increases the exchange current and reduces η_a. Values of i_0 vary from about 1 A/cm^2 to 10^{-80} A/cm^2: in most practical battery systems they are in the range 10^{-1} to 10^{-5} A/cm^2.

2.3.2 Concentration Polarisation

The flow of current at an electrode implies a reaction taking place with the concentrations of reagents being reduced and the concentrations of products being increased. The resulting concentration changes will alter electrode potential as indicated by the Nernst equation (2.3). Since the concentrations of the various species will alter with time, the value of concentration polarisation η_c will be time-dependent as well as dependent upon current density; the only exceptions being certain special electrode arrangements where a uniform flow of reactants to the electrode surface takes place (see Bard and Faulkner, 2.5).

Diffusion theory postulates that the layer of electrolyte in contact with the electrode surface becomes depleted of reactants, and the rate at which more reactant can transfer from the bulk electrolyte beyond this layer determines the ultimate rate of the electrode reaction. Transfer through the depleted layer is driven by diffusion and by electrolytic migration in the electric field between the electrodes. In the simple case with a single reactant limiting the reaction, the concentration polarisation is given by:

$$\eta_c = \frac{R.T}{n.F}.\ln(1 + 1/Y) \tag{2.12}$$

where

$$Y = \frac{a.n.F.D}{\delta}.(1 + n.t/z) \tag{2.13}$$

a is the bulk solution activity of the reactant involved.
D is the diffusion coefficient of the reactant in the relevant conditions.
δ is the width of the depleted layer at the electrode surface.
t is the transport number (describing the fraction of current carried by ionic migration in the electrostatic field).
z is the charge (ionic valency) of the reactant.

Equations 2.12 and 2.13 apply to the rather simplified case of a single reactant. Similar but more complex relations apply in the case of current controlled by both reactant and product diffusion. A special condition occurs when the reactant cannot diffuse fast enough to support the electrode reaction. In this so-called 'limiting current' condition, η_c increases abruptly. The value of current density required for this condition is given by:

$$i_l = \frac{n.F.D.a}{\delta} \tag{2.14}$$

The concentration polarisation η_c is then related to the limiting current, i_l, by

$$\eta_c = R.T/_{n.F}.(i_l/(i_l - i)) \tag{2.15}$$

Note that when the limiting current i_l is reached, concentration polarisation will change the electrode potential abruptly to values where an alternative electrode reaction can take place. In aqueous solution, this may well be to a positive value where oxygen evolution can take place or to a negative value where hydrogen evolution can take place, depending on whether the electrode is an anode or cathode. The current flowing through the electrode then becomes the sum of currents supported by the individual reactions at the particular electrode potential.

2.3.3 Resistance or 'Ohmic' Polarisation

Each of the cell components has a finite resistance and hence introduces a voltage drop when current flows. In galvanic cells, the electrolyte often introduces the highest resistive component. Porous separators are used to prevent the electrodes from touching and thus producing a short circuit. These separators allow current via ionic paths through their pores but still introduce a considerable restriction in the electrolyte path and hence increase the cell's internal resistance. Sometimes the active materials themselves have an inherently high resistance, and a supportive conducting network is required within the electrodes. Since resistive drop is determined by Ohm's law, resistance polarisation is proportional to current.

$$\eta_r = i.R \tag{2.16}$$

where R is the total internal resistance of the cell.

2.3.4 Total Voltage Change

The total voltage change which occurs when current is passed is the sum of the various polarisation effects.

$$E_0 - E = \eta + \eta_c + \eta_r \tag{2.17}$$

where E_0 is the equilibrium open circuit voltage of the cell, and E is the working voltage when the current is flowing.

Since η and η_c are both concentration-dependent, they both change with time unless ways of avoiding significant concentration changes are adopted. When the current through a cell is interrupted, η_r dies away instantaneously, whereas η and η_c decay gradually. Interrupting current in a regular and abrupt manner can therefore be used to resolve the resistive effects from other contributions.

There is a further effect which is related to concentration polarisation but is a special case and should be considered separately. This is the effect known as surface passivation. If an electrode reaction produces an insoluble product which is less dense than the original active material, this product may form a film which gradually obscures the electrode surface and prevents further reaction from taking place. The electrode may appear to have reached a limiting current which causes an abrupt change in potential. A related effect may occur as a reaction product gradually blocks the pores of a porous electrode, thereby preventing electrolyte access to the reacting surface. Such effects signal the end of the discharge in several familiar battery systems.

As noted before, polarisation effects ensure that cell voltages on discharge are always lower than cell voltages during charging, and therefore contribute to the inefficiency of a cell as a storage device. Since it is desirable for battery voltages to remain as constant as possible during discharge, it is important to reduce the polarisation effects by all practical means. In all cases, the polarisation effects are found to be a function of the current density at the electrode/electrolyte interface. It follows that electrodes with a large surface area will be able to support larger currents for a given polarisation-voltage drop. As a result, battery engineers have devised various means of achieving the largest surface areas possible for a given amount of active material. In the case of the lead-acid cell, the active materials are produced in highly porous forms so that the total surface exposed to electrolyte is many times larger than the apparent surface area (surface areas of tens of square metres per gram are achieved for lead electrodes and several square metres per gram for lead dioxide electrodes). Not all of this surface will be available for reaction, and resistive-drops in the electrolyte down the pores limit effectiveness. Nevertheless, the porous electrodes prepared by the Fauré method of electro-reducing or oxidising pastes of lead sulphate provide electrodes of vastly superior performance to older types which used solid materials on lead sheet (see Chapters 1 and 3).

The activation polarisation effects are, as stated, a function of the particular electrode reactions and can only be materially affected by changing the chemical system. However, any agent which increases the rate of the electrode reaction will reduce the activation polarisation losses. In practice, increased operating temperatures, catalysts and increased operating pressures are used to achieve this end in other types of power-producing cells. Such effects are not generally accessible for the lead-acid cell, but it is favoured by having inherently fast electrode reactions (exchange currents of the order of 10^{-3} amps/cm^2).

Concentration-polarisation effects can be diminished by providing improved transport of material to the reaction surface. This could be achieved by stirring the electrolyte (actually used in some special purpose lead-acid batteries) or by designing cells to allow natural convection and diffusion effects to be most effective.

Resistive effects can be diminished by proper dimensions of electrode and cell connections so as to minimise the voltage drop between the site of the reaction on the electrode surface and the cell terminal. In the lead-acid cell, a large voltage drop occurs due to electrolyte resistivity; the electrodes are therefore placed as close together as possible and are separated by porous materials with the minimum of obstruction to the ionic current flow in solution.

2.3.5 Polarisation Limitations on Effective Cell Capacity
Cells and batteries are normally rated in terms of the charge (ampere-hours,

abbreviated to Ah) they can supply from a fully charged state until the voltage drops to a prescribed limit. It is obvious from equation 2.15 and the above discussion, that the higher the current density and the higher the total polarisation-voltage drop, then the sooner the discharge voltage limit will be reached and the lower will be the effective cell capacity. Depending on the currents involved and the nature of the electrode materials, the limit may be reached well before all the active material is actually consumed. This is indeed the case with the normal lead-acid cell; and in typical automotive battery designs, only 30 to 50% of the active material on the electrodes may be used before the discharge voltage limit is reached (in a 20h-rated discharge). The actual values will depend on the relative amounts of all the materials involved and the practical voltage limits chosen, which must be a compromise between the application demands and the economics of cell construction.

The fall in effective capacity with increasing current was recognised experimentally at an early stage and an empirical equation, the 'Peukert Equation' (Peukert, 2.7), was devised to relate effective capacities at various current densities.

$$K = t.i^n \qquad\qquad (2.18)$$

where K is a constant representing rated capacity,
t is the time to reach a discharge voltage limit,
i is the steady discharge current,
n is a constant (of numerical value greater than 1) related to cell design.

The Peukert equation holds tolerably well over a restricted range of currents and for relatively steady discharges. If the discharges are intermittent or of widely varying current values, it becomes inaccurate. Concentration-polarisation values in particular will vary continuously in the latter type of discharges and the voltage limit will be reached in a different way to that occurring in a steady discharge. It is a well known fact that a battery which has apparently reached its discharge limit at a high current will often recover when left on open-circuit and will provide additional discharge capacity at a lower current. During the period on open-circuit, the concentration gradients built up at the cell electrodes relax, and reduce the limiting polarisation effects.

The effect of concentration gradients built up at the electrodes is also observed on recharging the cell. The concentration changes built up on discharge actually assist the charging reactions. A cell will therefore absorb charge current at a lower voltage if the charging occurs immediately after a discharge. If the cell is left to equilibrate for any length of time, the same charge current will produce a larger increase in cell voltage due to the concentration gradients relaxing. In a similar manner, a cell will give a higher discharge voltage immediately after charge as compared to a similarly charged cell that has been allowed to equilibrate for some time.

These time-dependent concentration-polarisation effects complicate the modelling of battery behaviour (see Chapter 5) but in the case of automotive batteries, which are continually switching from charge to discharge, they work to the benefit of the overall efficiency of the electrical system.

2.3.6 Thermal Effects in Working Cells

In practice, virtually all cells are observed to warm up on both charge and discharge. In theory, heat evolution is determined by two factors: the entropy change of the cell reaction, and the losses due to polarisation effects. If the entropy change of the cell reaction is positive, heat is released to the surroundings and if the entropy change is negative, heat is absorbed from the surroundings and the cell cools down.

These thermodynamic effects are usually overwhelmed by the heating effects due to polarisation losses. The heat evolved in this case is equal to the difference between the equilibrium cell voltage (i.e. at zero current) and the actual working potential multiplied by the total current flowing. Thus the overall heat evolution is given by:

$$\Delta Q = I.(^{T.\Delta S}/_{n.F} + \eta) \tag{2.19}$$

where ΔQ is the total rate of heat evolution at a cell current I
and η is the total voltage change due to polarisation losses.

It is possible to define a 'Thermoneutral Potential', E_{tp}, where the polarisation losses just balance the thermodynamic effects and no heat is passed to or from the surroundings.

$$E_{tp} = {}^{\Delta H}/_{n.F} \tag{2.20}$$

2.4 The Lead-Acid Cell

The above generalisations lead us to the special case of the lead-acid cell which forms the basis of the vast majority of automotive batteries. As noted in Chapter 1, the original lead-acid cells were made by Planté using electrodes of solid lead with the active materials formed by prolonged electrolysis. The actual active materials in a fully charged cell are lead dioxide on the positive electrode, and metallic lead on the negative electrode. The sulphuric acid which is used to form a conducting electrolyte between the electrodes also takes part in the cell reaction and forms the third active material. The cell reactions are shown schematically in Figure 2.4.

Several important points arise from the reactions shown in Figure 2.4. Firstly, as noted earlier, the sulphuric acid in the electrolyte takes part in the reaction. Thus the electrolyte solution becomes more dilute during discharge and this fact has frequently been used to indicate the battery's state of charge (usually by measuring concentration in terms of specific-gravity). The concentration change also produces a reduction in conductivity which contributes to a lowering of power-output at low states of charge.

Lead sulphate is produced at both electrodes. This is only slightly soluble in sulphuric acid and forms a layer which gradually obscures the electrode surfaces (cf. passivation as described in Section 2.3.4). In practical electrode structures, less than 50% of the active material may be used before blocking by the sulphate layer and associated concentration-polarisation effects reduce the cell voltage below its discharge limit. The sulphate also has a much lower density than lead or lead dioxide. and the resulting volume changes produce stresses in the electrode structures which gradually cause disintegration and loss of contact between the active materials and the conducting base of the electrode. The detailed properties of the active materials and the cell reaction characteristics are found in Appendix 2.

On Discharge (Reversed for Charge)

Positive $\qquad PbO_2 + 4H^+ + SO_4^{2-} + 2e \rightarrow PbSO_4 + 2H_2O$

Negative $\qquad\qquad Pb + SO_4^{2-} -2e \rightarrow PbSO_4$

Overall $\qquad \mathbf{PbO_2 + Pb + 2H_2SO_4 \rightarrow 2PbSO_4 + 2H_2O}$

On Overcharge

Positive $\qquad\qquad 2H_2O - 4e \rightarrow O_2 + 4H^+$

Negative $\qquad\qquad 4H^+ + 4e \rightarrow 2H_2$

Overall $\qquad\qquad \mathbf{2H_2O \rightarrow 2H_2 + O_2}$

Fig.2.4. Lead-acid cell reactions

It may be noticed from the individual electrode potentials that both electrodes are thermodynamically unstable in water. The lead electrode operates below the hydrogen electrode potential and thus lead should reduce water to hydrogen. Similarly the lead dioxide electrode potential is above the oxygen electrode value and lead dioxide should oxidise water to oxygen. The fact that the electrode materials enjoy a remarkable level of stability is due to the slowness of the hydrogen and oxygen evolution reactions on lead and lead dioxide surfaces respectively. Impurities which are deposited on the electrodes, and which support the gas evolution reactions at higher rates (e.g. iron, nickel), cause self-discharge of the electrodes and deterioration of the cell. One common source of such an impurity which allows gas evolution is antimony, the alloying metal commonly used to impart extra strength to the plate-support grids. As batteries age, the positive grids in particular corrode, and release antimony-containing ions into solution. These are reduced at the negative electrode surfaces to provide sites for hydrogen evolution at higher potentials than would occur with a pure lead surface. The resulting extra hydrogen evolution provides a cathodic process which self-discharges the lead electrode. It also involves a loss of current for the charging process and thereby degrades performance and shortens the electrode life.

As indicated in Figure 2.4, when the electrode potentials are pushed beyond the normal values, as in overcharge, gas evolution at significant rates will occur. This gas evolution leads to a loss of water from the electrolyte which needs to be replenished. This is discussed further in Section 2.5.

2.4.1 Electrode (Plate) Structures

It was noted in Section 2.3.4 that polarisation effects depended on current density rather than absolute current values. The objective in making effective battery electrodes is thus to achieve the highest surface-area possible for a given weight of material. In lead-acid batteries this is achieved by the use of the Fauré process. Fauré (2.8) found that pastes of lead oxides and sulphuric acid formed mixtures which could be coated onto suitable conductive grid networks and then formed electrolytically into the active materials. Similar mixtures could be used for both electrodes, and the final active materials were determined by the electrode treatment in the formation process.

The commercial process is described in detail in Chapter 3. Metallic lead is first oxidised to form litharge, PbO. This is achieved either by the attrition of lead-shot in a ball mill with controlled amounts of water added, or by direct oxidation of molten lead. The resulting powders contain large amounts of particulate lead, which is gradually oxidised in later stages of preparation. Experience has shown that this type of material produces a better result than the use of pure litharge.

The leady oxide produced is then mixed with water and sulphuric acid to produce a paste, containing unchanged lead, lead oxide and a variety of lead sulphates with a general formula $PbSO_4.XPbO$ where X varies from 0 to 4. The nature of the sulphate mixture has a profound effect on the performance and life of the resulting active materials (Bode, 2.9), since the volume-changes produced in later reactions determine the porosity of the active material, and its crystalline structure influences the mechanical strength and service life.

Paste mixtures produced in this way are then spread onto lattice support grids which give mechanical support and provide an electrically conducting path to the cell terminal. Once pasted, the grids are allowed to stand in temperature and humidity controlled environments for periods up to several days. During this 'curing' process, the grid surfaces are corroded in a way that forms an adhesive bond with the pasted material. Surface layers of oxide and sulphate are formed on the grid surface which 'key' into the paste structure, and the paste itself warms up due to the chemical reactions taking place, with the result that it hardens into a cement-like mass. The unchanged lead added as part of the original litharge is oxidised further and the cured paste contains a variable mixture of monobasic lead sulphate, $PbO.PbSO_4$, the tribasic sulphate, $3PbO.PbSO_4.H_2O$, and some tetrabasic sulphate, $4PbO.PbSO_4$.

The actual active masses are produced when the cured pasted grids are placed in sulphuric acid electrolyte (either separately or assembled into cells). In the formation process, the sulphate mixtures are converted to lead dioxide, PbO_2, on positive grids and to metallic lead on negative grids, in reactions similar to the charging reactions shown in Figure 2.4. The electrodes or 'plates' so formed have a high internal porosity and consequently a very high effective surface-area (several square metres per gram). The volume changes produced are due to the large difference in molar volumes between the starting materials and the final active materials which are the products as shown in Table 2.1. It can be seen that since one molecule of lead sulphate produces one molecule of lead or one molecule of lead dioxide, very large volume changes occur in the material structure, resulting in high porosity and surface area.

Note that two different crystal types of lead dioxide exist: the α orthorhombic form and the β tetragonal form. The relative amounts formed depend on specific processing conditions, and the precise nature of their influence on the life and performance of the positive plate is still a matter of research. However, there is evidence to show that the α/β ratio influences the structural integrity and durability of the active material (Dasoyan and Aguf 2.10; Burbank et al, 2.11).

The pastes used to form both positive and negative plates could be the same, but experience has shown that certain additives have a beneficial effect on plate life and performance, and differ in their effect between positive and negative plate. As a result, it is commercial practice to add to the pastes intended for negative plates small (ca. 1%wt) amounts of barium sulphate, lignosulphonates, and carbon black. These

Table 2.1. Range of density and volume for the plate materials

	Molecular Weight	Density g/cm^3	Molar Volume cm^3/mole
PbO	223.2	9.35	23.9
PbSO$_4$	303.3	6.3	48.2
PbSO$_4$.3PbO	990.8	7.0	142
PbSO$_4$.4PbO	1196.0	8.1	149
Pb metal	207.2	11.3	18.25
αPbO$_2$	239.2	9.8	24.3
βPbO$_2$	239.2	9.5	25.2

are sometimes referred to as 'expanders' since they were originally thought to expand the plate's surface area. The effect on the negative plate is to improve performance at low temperatures and to lengthen the charge-discharge cycle life. Actual expander mechanisms are a subject of both experiment and speculation (Hoffman and Vielstich, 2.12; Szara, 2.13) but the barium sulphate is thought to provide a crystal nucleus for lead sulphate formation and to prevent surface blockage by this material. The carbon black may influence conductivity in the formation stages, but its exact function is a mystery.

Lignosulphonates are produced by extracting certain woody materials (originally the cell's wooden separators) with sulphuric acid. They normally appear as brown powders of poorly defined chemical composition. Their action may be due to the absorption of the material on growing lead sulphate crystals, thereby changing their shape and preventing the formation of a dense impervious layer, and it may also be due to modification of the lead surface produced when the plate is formed or recharged.

In recent years, other additives have been used with both positive and negative plates. These have included organic and inorganic fibres to strengthen the active mass, and binders based on fluorocarbons to maintain the integrity of the active materials. Few have been used on the scale of the negative additives referred to above.

Figure 2.5 shows the stages in the preparation of battery plates and in the construction of complete cells. The figure also shows one important component that has not yet been described. This is the separator used to prevent the plates from coming into electrical contact and short-circuiting the cell. This particular component must be electronically-insulating but must provide an unimpeded ionic path between the plates. Traditionally this is achieved by the use of a thin lamina of highly-porous material. Originally special acid-resistant woods were used (these, incidentally, provided the first source of the lignosulphonate expanders referred to earlier). However a range of specially-prepared proprietary microporous materials based on polythene, PVC and other common polymers are now available and provide a reduced ionic resistance together with better chemical stability. Ideally the separator should be as porous as possible and as thin as possible to provide the lowest resistance between the plates.

The plate support grids are of special importance. They provide the major current path from the site of the reaction on the electrode surface to the cell terminal. Pure lead would be an ideal material but it is mechanically weak. A variety of alloying elements such as antimony, bismuth, tin, and calcium may be used to strengthen lead. The traditional alloy used in batteries has been lead-antimony (up to 10% Sb). However, the antimony is leached into solution and provides sites on the negative electrode where hydrogen evolution can occur at significant rates and constitute a self-discharge process. Lead-calcium (sometimes with tin) is now used extensively in spite of handling and casting difficulties, and this eliminates the effects seen with antimony. It may, however, form resistive surface-corrosion layers on its surface when plates are deeply discharged.

2.4.2 The Functional Components of a Lead-Acid Cell

It is appropriate to review the functions of a practical cell's working parts as follows.

Active Materials - To provide the positive and negative electrode reactions. The materials should have the highest surface-area possible in contact with the electrolyte so as to minimise polarisation effects. In addition, the materials should have stable structures with the lowest possible electrical resistance.

Support Grids - To provide a mechanical support for the active materials and to provide a low-resistance current path to the cell terminals. The grids should be mechanically strong, of high conductivity and should be corrosion resistant.

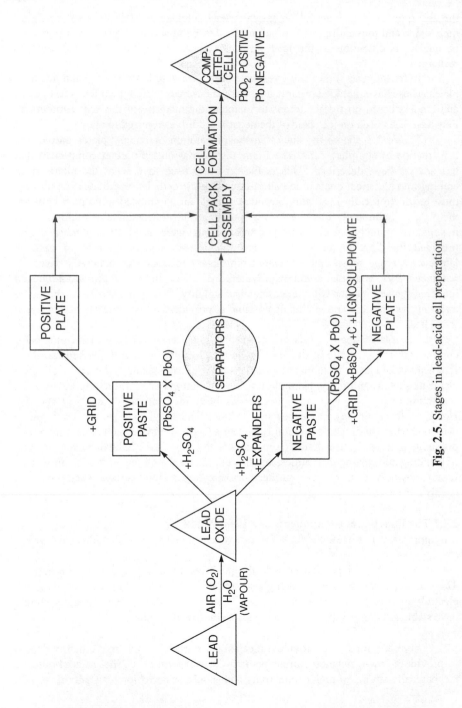

Fig. 2.5. Stages in lead-acid cell preparation

Separators - To prevent electrical contact between the plates and to present the lowest possible resistance to ion-flow through the electrolyte between the plates.

Electrolyte - To provide the ionic path between the plates, and in the lead-acid cell to provide a source of sulphuric acid as one of the active materials.

Cell Terminals - To provide a low-resistance electrical connection to the outside world. The terminals may be connected to internal cell busbars linking all positive plates to one terminal and all negative plates to the other terminal.

Cell Container - A suitable chemically-resistant, insulating container which encloses the cell components. It normally has vents to allow escape of gases formed on overcharge, and to allow filling with electrolyte.

The traditional cell structure has plates of alternating polarities interwoven with separators to allow full use of both electrode surfaces. An alternative structure uses a Swiss-roll of positive and negative plates with a separator in between. This, however, gives only cylindrical cells. Another structure, the so-called bipolar battery, has composite plates of dual polarity stacked with separators to form a pile. (This is the original structure used by Volta.) It is difficult to implement in lead-acid chemistry since a corrosion-free layer is required between the composite electrode surfaces. All the structures are illustrated in Figure 2.6.

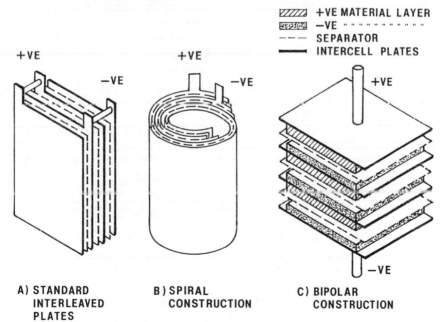

Fig.2.6. Alternative cell constructions

2.5 Water Loss and Maintenance-Free Cells

As mentioned in Section 2.4, polarisation effects move the electrode potentials intoregions where hydrogen and oxygen are evolved from the lead (negative) and lead dioxide (positive) electrodes respectively. Indeed, one common indication of a fully charged cell is free gas evolution.

The onset of gas evolution is not a sharply defined phenomenon. What actually happens is that a gradually increasing amount of charging current goes into oxygen evolution at the positive plate, and into hydrogen evolution at the negative plate. The processes at the positive and negative do not necessarily match each other, and the gas evolved at the start of overcharge is normally rich in oxygen. As both electrodes reach a fully charged state, the gas composition gradually moves towards the stoichiometric value (i.e. $2H_2+O_2$). The amount of gas evolved and the exact composition are a function of current density, state of charge and temperature (Webley, 2.14; Peters, 2.15). Very low temperatures (below -10°C) and high current densities favour a large proportion of the current going to gas evolution even before full-charge is reached.

When both electrodes are charged, it might be assumed that all of the cell current goes into gas evolution. However it is observed (Peters, 2.15) that a small fraction is not used in gas evolution and may be used in the formation of highly oxidised products such as persulphates and peroxides at the positive electrode:

$$2SO_4{}^{2-} - 2e \rightarrow S_2O_8{}^{2-}$$

$$2H_2O - 2e \rightarrow H_2O_2 + 2H^+$$

These reactions are of course matched by equal currents going into hydrogen evolution at the negative electrode. Batteries that have been overcharged for some time will sometimes exhibit anomalously high open-circuit voltages due to the presence of these compounds. They may persist for some days unless a discharge takes place.

The decomposition of the electrolyte to produce gas is obviously undesirable since:

a) The electrolyte acid becomes more concentrated and grid corrosion effects and chemical attack on the separators are accelerated,

b) The electrolyte level may fall, exposing the plates and reducing the cross-section available for conduction. Plate exposure to air may also accelerate self-discharge effects and irreversible capacity loss,

c) The gas evolved is a flammable mixture and presents a significant explosive hazard. (The gas will also be saturated with water vapour and may carry droplets which represent additional water losses to those caused by electrolysis.)

Water loss and the resulting topping-up requirements can be reduced by operating the cells at voltages below those required to cause gas evolution in normal operation. Most automotive electrical systems are therefore voltage-limited (see Chapter 1) and are limited at 14 to 14.5 V for a nominal 12 V battery. The exact voltage chosen is a compromise between water loss/gas evolution and rate of recharge in the particular system duty-cycle. Too high a voltage causes excessive water loss and too low a value prevents full recharging in the operating time available.

Ideally, the voltages are reduced as the temperature rises, since polarisation effects decrease with temperature, and a given voltage will force a larger current into the battery. A negative temperature coefficient of about -10 mV/°C is applied in some cases. As with the voltage limit, the exact value is chosen to match the particular duty cycle and temperature-range experienced.

2.5.1 Reduced Maintenance Batteries

Since the only significant maintenance requirement is electrolyte water replacement, some means of preventing electrolytic water loss is required. As noted above, water electrolysis can be minimised by limiting cell charging voltages, but this can still allow considerable gas evolution in higher states of charge. There are three design strategies which have been used to reduce water loss to very small levels in normal automotive (voltage limited) operation. The first uses special electrode-support grids to slow down the water electrolysis reactions, and the others rely on recombining gas evolved to reform the water and return it to the electrolyte.

a) Grid Alloys with Reduced Gas Evolution

It was noted earlier that gas evolution rates depended on the properties of the electrode surfaces where the electrolysis reaction took place. The original alloys used for support grids could contain 10% or more of antimony. When corrosion took place, this would be electro-deposited on the lead surface of the negative and would provide sites for fairly rapid hydrogen evolution. If alloys with other metals such as calcium are used, there are no deposits formed on the negative plates which act as hydrogen evolution sites. Gas evolution and resultant water loss are then substantially reduced and, with the aid of an extra electrolyte capacity, batteries used with correct voltage limits do not need topping up in their expected lifetime.

b) Catalytic Recombination of Evolved Gases

The hydrogen and oxygen evolved on overcharge can be recombined in stoichiometric proportions using a suitable catalyst. Pellets of high surface-area alumina coated with platinum or palladium have been inserted into the tops of cells (mounted clear of the electrolyte) for this purpose. Water vapour is formed which can condense and fall back into the electrolyte. The catalyst and support must be designed in such a way as to avoid being flooded by condensed water. In addition the catalyst must not rise to such a temperature that the gas mixture is ignited with a resulting explosion. A further problem is the poisoning of the catalyst by some gases generated within the cell, particularly as ageing takes place. These may include stibine (SbH_3) from grid corrosion, or sulphur dioxide from separator degradation. The biggest disadvantage of catalytic recombination is that the oxygen and hydrogen are not always evolved in the stoichiometric quantities for water formation. In general, oxygen is evolved first and in excess quantities. It is therefore not possible to close the cells and some imbalance of materials gradually occurs. Although catalytic recombination has been used successfully in other types of battery, it has only met with limited success in lead-acid cells.

c) Oxygen Recombination Cycle

It has been mentioned earlier that oxygen is the first gas to be evolved as the battery reaches full charge. It is possible to adjust the balance of active materials so that hydrogen is not evolved from the negative until long after the positives are fully charged and substantial overcharge has taken place. If the initial oxygen evolved has good access to the surface of the lead negative electrodes, it is recombined with the resulting formation of lead sulphate according to the overall reaction

$$O_2 + 2Pb + 2H_2SO_4 \rightarrow 2PbSO_4 + 2H_2O$$

As further charging takes place, the lead sulphate formed is reduced back to lead and the cycle is repeated. Provided the overcharge current is kept within limits, this cycle can contain the oxygen formed without any significant rise in pressure (Harrison and Wittey, 2.16). The cycle inhibits hydrogen evolution from the negative and the cell can therefore be virtually sealed (Thompson and Warrel, 2.17). Valve regulated batteries relying on this strategy are described further in Section 2.13.

2.6 Battery Design Considerations

A definition of the vehicle, be it an off-road construction vehicle, a passenger-carrying coach, a private car etc., which states its required performance, its range of usage patterns, and environmental requirements, is clearly the first stage in developing a suitable battery design for that vehicle.

Indications of potential specific-application variants, for example police work or ambulance work, are also important, as indeed is the range of variations of engine size and equipment levels to be considered.

In a vehicle the battery is called on to fulfil two main functions: firstly to rotate the flywheel/crankshaft assembly of the engine at a speed sufficient to get the engine started, and secondly to power the other items of electrical equipment on the vehicle. In order to carry out these functions over a useful lifetime the battery must be able to withstand the various mechanical stresses to which the vehicle environment subjects it. As an illustration of these sorts of stresses, forces greater than 10 G, due to vibration in underbonnet locations, have been measured on batteries on cars being driven over rough roads, such as French pavés.

2.6.1 Electrical Function

Vehicle starting. The average family car engine can require 2 to 3kW of power in order to revolve the crankshaft assembly at a rate sufficient to achieve starting. Figure

2.7 shows the current draw needed to start a 2.5 litre petrol engine at normal temperature, while Figure 2.8 provides the power curve for an SLI battery used for starting. The precise requirement depends on the characteristics of the engine, with the number of cylinders, compression ratio, diesel or petrol versions, heading the list. Information regarding the type of starter-motor, the gearing of the starter pinion and ring gear and design of the flywheel come next. In addition, for a petrol engine, starting is vitally dependent on the high-tension circuit of the coil, distributor and spark-plug, the power for which is also supplied by the battery. In order to carry the starting current of several hundred amps., cables and terminations need to be of large cross-section, and the distance between battery and starter motor as short as possible.

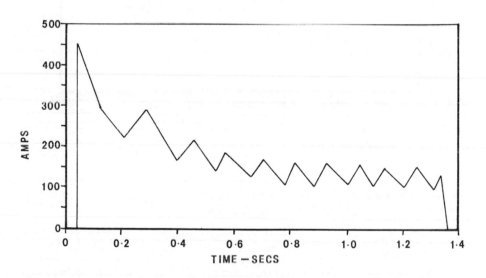

Fig.2.7. Engine starting current draw for a typical 2.5 litre petrol engine
(temp. approx. 15°C)

The temperature-range over which starting is required is a vital consideration. At very low temperatures oil is very viscous, engine resistance is high and the battery performs at its worst. Vehicle manufacturers carry out a great deal of testing of new engines and associated equipment in cold-room situations to resolve potential difficulties. The normal minimum temperature specified is -18°C but for some climates -29°C is specified. In extreme conditions battery heating can be required, or batteries are physically removed from the vehicle and kept warm.

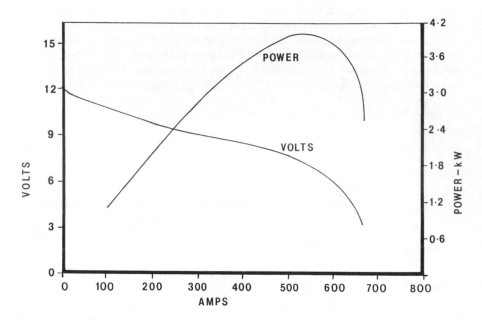

Fig.2.8. Battery power curve for 70 Ah (C_{20}) battery at 25°C

Other loads. These need to be examined under three operating conditions: engine off, engine running normally, and engine running with charging failure.

Engine off, or parked condition. These loads are quite small, consisting of items such as clocks, alarm systems and computer memories, and can be measured in milliamps. However, even such small currents can, over a period of time, be sufficient to prevent starting. A battery capacity of about 50% of its specified design value is normally considered to be the minimum level for starting an engine in average condition under adverse circumstances. Given that some devices can consume up to 50mA this situation can be reached with a battery of nominally 40Ah in as little as three weeks. Cars are not infrequently left parked at airports for this length of time!

Engine running normally. Modern alternators are capable of producing useful current even at engine tickover speeds, while at normal operating speeds they can provide a large proportion of the maximum anticipated loads, including headlights, heater-fan, rearscreen heater, radio and others simutaneously. Only at low engine speeds and exceptional load conditions is the battery called on to make good a deficiency, remembering of course that in fact the alternator output is used for charging the battery which in turn sustains the load. Under such conditions the battery operates in a similar way to that of a smoothing capacitor, preventing the large fluctuations in alternator output, as the engine speed changes, from affecting the performance of the various devices in use.

Engine running with charging failure. This clearly is an emergency situation when the fan belt driving the alternator pulley fails, or a defect arises in the alternator circuitry. Under these conditions the battery must be able to sustain all essential loads to keep the vehicle mobile safely for a limited period until emergency repairs can be effected.

2.6.2 Mechanical Function

Size, weight, and shape are key parameters influencing the position and installation of the battery. Size and weight are determined to a large extent by the electrical requirements specified. Some latitude is available regarding shape, although even here, with current technology and production methods, only rectangular, prismatic forms are generally considered.

Temperature resistance, vibration resistance and shock resistance are all important mechanical parameters influencing the designer, as are decisions regarding location, be it in the engine compartment, with proximity to heat emitters such as engine or exhaust manifold, in the cabin, with potential problems of gas emissions, or in the luggage compartment, with likely problems of long cable-lengths to the starter motor.

Most batteries contain free electrolyte which can leak out through gas exits or filling holes if tilt angles are excessive, or if the battery is inadvertently overfilled during topping-up. (Charging increases the volume of the electrolyte. See Section 5.5.2.) Hence, as the electrolyte in a lead-acid battery is highly corrosive sulphuric acid, proximity of the battery to sensitive components such as brake pipes should be avoided at all costs.

The possibility of dust or water ingress is another aspect that needs to be considered. Dust can clog gas exits and cause covers on filling apertures not to seal correctly. Layers of dirt or dust between battery connectors can absorb moisture, become conducting with the small amount of sulphate or sulphuric acid often present on the lid of the battery and cause battery discharge. Water ingress can cause the electrolyte to become weaker and can lead to flooding, causing the escape of an acid solution with subsequent problems of vehicle corrosion and/or loss of battery performance.

Fixing of the battery to the vehicle is also a key consideration. The installation engineer is dealing with a relatively dense and massive component that contains electrochemical energy and corrosive fluid. It must be fixed in a manner that prevents other metallic parts of the vehicle from causing short circuits with the terminals and prevents the battery from coming loose under severe vibration or shock conditions, including accidental impact.

However, the location and fixing need to be such as to make installation and removal relatively easy, and to make in-situ inspection and maintenance simple.

Fig.2.9. A typical modern cell pack
(Lucas Yuasa Batteries Ltd)

2.7 Battery Internal Design

The basic architecture of the lead-acid battery has remained largely unchanged for many years, consisting of alternating positive and negative thin rectangular plates interleaved with separators comprising a cell, as shown in Figure 2.9, with each cell in adjacent compartments of a prismatic box, interconnected with top parts of metallic lead alloy external, or, more recently, internal to the sealing covers.

However, within this framework many changes have taken place as a result of the development of new materials, better processing and manufacturing technology, and a better understanding of the way in which a battery functions.

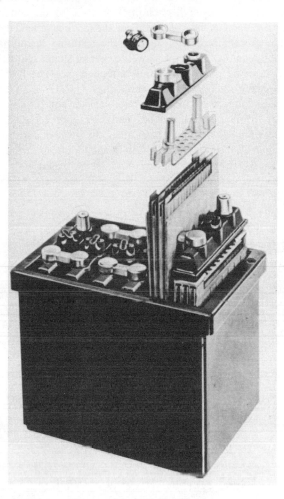

Fig.2.10. A battery construction from c.1935
(Lucas Yuasa Batteries Ltd)

Figure 2.10 shows a battery typical of those produced in the 1930s. The multiplicity of components and complexity of intercell connector and lid sealing can clearly be seen. One of the many weaknesses of such designs was the difficulty in achieving adequate sealing. However, such designs can still be seen in some countries in the developing world. Indeed they are often preferred to the more powerful modern designs for they have one important attribute, namely, ease of repair. External connections can be cut off or melted off, the pitch can be melted out, the cell pack can be lifted out and repaired with perhaps a single new plate or separator, replaced, and the whole encased in a new layer of pitch prior to re-fusing the connector bar. These activities are frowned upon as they often expose the repairer to lead dust or fume with inadequate or non-existent protection against the resulting health-hazard, which is described in greater detail in Chapter 4. However, repair shops, using these techniques can still be seen in some poorer countries, although the total number of batteries suitable for such treatment is continually decreasing.

The gradual evolution of the battery from this design through to the slimmer, sleeker designs of the late 1960s can be seen in Figure 2.11. The most obvious changes are the re-siting of the connector under the lid and the replacement of the single cell lids with a single-piece 'multi-lid'. The later change from this inset pitch-sealed 'multi-lid' to the stronger 'wrap-over' lid design sealed with epoxy resins can also be seen.

Fig. 2.11. Battery evolution from 1935 to 1970
(Lucas Yuasa Batteries Ltd)

Figure 2.12 is of a modern 1990s battery, using polypropylene for the lid and container, the basic technology for which was developed in the mid 1960s. The evolution of the interconnector through these stages can be seen diagrammatically in

Figure 2.13. As a result the modern battery is smaller, lighter, more reliable, longer-lived, requires less maintenance and is significantly cheaper in real terms than its forebears. It has also become more functionally specialised and to some extent has given the battery designer a greater ability to tailor-make a solution to a specific need.

2.7.1 General Design Trends

Until the mid to late 1960s battery plates were of the order of 2 to 2.5mm thick. The grid had to be very substantial to make up for casting defects inevitably present with casting technology as it then was, to withstand handling within the factory, and to withstand the high levels of overcharge suffered by the battery from the quite crude charging control systems then available.

As it was, one of the main reasons restricting battery life to about 2 years was structural failure of the positive grid through overcharge corrosion, failure occurring in areas of weakness caused by casting defects. As areas of the plate became electrically detached from the rest so the current density on the rest of the plate would increase, thus accelerating the eventual failure of the battery.

Fig.2.12. A modern car battery
(Lucas Yuasa Batteries Ltd)

46

Paste weights are generally a function of grid thickness. Therefore as the grids of the time were thick, so paste weights were high. This meant that relatively high low-rate capacities could be achieved from a small number of plates, but the high-rate discharge capability of a battery of any particular size was limited. Hence, in order to achieve higher levels of starting performance, physically larger and heavier sizes of battery needed to be specified. The problem was compounded by the use up to that time of battery containers made of rubber/pitch/coal dust mixtures which in themselves were of a thick, heavy, construction necessary to withstand the mechanical stresses experienced.

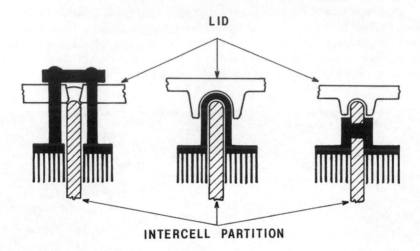

Fig.2.13. Schematic showing size and path length of loop-over connector vs TPW (through-partition welding)

Three developments of the late 1960s were key to the way in which battery design flexibility has improved:

1) The development of the semi-conductor, which led to the introduction of solid state regulators, and to the birth of the power diode, enabling the use of alternator type generators to be seriously considered for cars.
2) The development of polypropylene copolymers which transformed the battery container, shrinking overall battery size and weight and overcoming many problems of leakage and inadequate structural integrity associated with rubber containers.
3) The development of better lead alloys and improved casting technology. Defects were largely eliminated from the clipped grid, thinner sections were able to be cast, and thinner and lighter grids were made, of handleable strength.

Table 2.2 illustrates the way in which constituent parts of the battery have changed between the battery shown in Figure 2.10 and that shown in Figure 2.12. Energy-density has increased from 19Wh/kg to 36Wh/kg while power-density has increased from 57W/kg to nearly 200W/kg in some of the latest designs.

Packaging, that is, the container/lid assembly, accounts for the biggest change, mainly resulting from the introduction of polypropylene, but the other constituents have also played their part.

Further development of grid manufacturing and handling technology have brought us to the present-day position where plates of the order of 0.8mm in thickness are available. The battery cell can therefore contain many more plates for the same active material mass, thus dramatically improving high-rate discharge capability of the battery where necessary. This means that, within constraints imposed by production economics, batteries can be tailored to provide the best compromise between low-rate capacity and high-rate capability.

Table 2.2. Change in component % weight over 55 years

Weight	1935 40Ah(20hr) 21kg	1990 40Ah(20hr) 11kg
	%	%
Packaging	25	7
Separators	0.5	5
Top Lead	10	4
Grid Lead	17	18
Acid	28.5	40
Active Material	19	26

Plate enveloping technology has enabled a further reduction in size to take place. Edge and bottom shorting has been largely eliminated by sealing each plate of one polarity type in its own separator envelope. The bottom reservoir under the plates, there to capture material eroded from the surface of the plates, (commonly known as the sludge-space) has been eliminated. Compare Figures 2.14 and 2.15

The introduction of the polypropylene container led to the development of cell-to-cell interconnections through the cell partition rather than over the top of the partition necessary with rubber containers. Although even today this connection is made above the level of the electrolyte to avoid the remote possibility of electrolyte leaking from cell to cell (which leads to self-discharge) designers continually strive to make the conducting path-length from cell to cell as short as possible. Even within the cell the route from the grid current-collector up through the bus-bar connecting all the plates of the same polarity and on to the interconnector is made as short as possible to avoid voltage drop and of course to reduce the amount of inactive and unrequired lead used. Modern battery alloys, together with the use of other more 'pure' components, have given rise to batteries that, when used with efficient reliable charging control

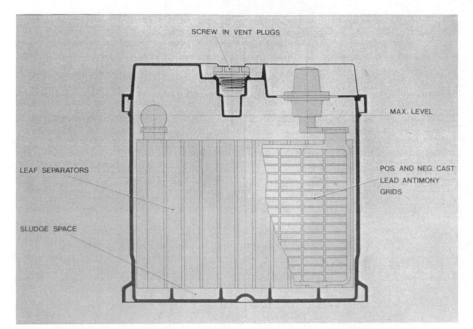

Fig.2.14. Cross-section through a modern standard battery

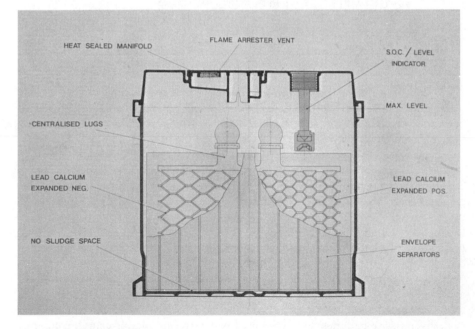

Fig.2.15. Section through a modern enveloped-plate battery

systems, require little or no maintenance throughout the life of the battery. Older alloys containing 6% or more antimony typically required topping up every 20,000 miles or so. Alloys containing lower levels of antimony, in the range 1 to 3%, extended this to about 50,000 miles, whilst the latest antimony-free alloys can give well in excess of 100,000 miles before the electrolyte level falls below the top of the plates.

Hence the requirement for easy methods of 'topping-up' the electrolyte with water at regular intervals has been largely eliminated. The vent plugs or manifolds have been replaced by non-removable plugs, often breathing through explosion-proof plastic-sinter discs. Another benefit of the newer alloys and 'cleaner' chemistry is that self-discharge is much reduced. Taking the same alloy groups as above, the charge retention after 12 months' stand, under the same conditions would be less than 25%(>6%Sb), 50%(1-3%Sb) and greater than 60%(0%Sb) respectively.

The latest trend, after a faltering start in the mid-1980s, has been the development of the valve regulated battery (VRLA - Valve Regulated Lead Acid), with a gelling agent incorporated in the electrolyte or the use of ultra-fine fibre-glass separator material with limited absorbed electrolyte. Such batteries can offer a good degree of gas recombination, and can be sealed to all intents and purposes. This subject will be expanded upon under Section 2.12 later in this chapter.

2.7.2 Detailed Plate and Cell Design
The conventional cast grid consists of a rectangular lattice-work with vertical ribs, horizontal strands, a surrounding frame, a lug on the top part of the frame and two small extensions or 'feet' on the bottom part of the frame. Figure 2.16 shows typical modern cast positive and negative grids.

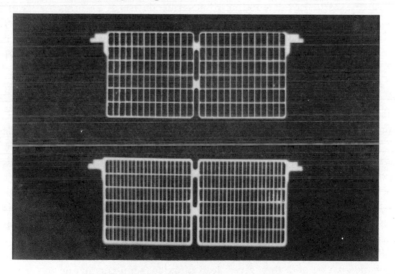

Fig.2.16. Photograph of typical cast positive and negative grids
(Lucas Yuasa Batteries Ltd)

The horizontal strands are normally the thinnest in cross-section but the largest in number, usually of a basic triangular cross-section, and cut alternately into one or other of the two halves of the moulding tool. Ribs are of a diamond cross-section, while the frame, with a lozenge shaped cross-section, has the largest cross-section. Grid thickness is defined by the thickness of the frame, with ribs and strands being set in by a small amount. The position of the lug is defined by the design of the connecting strap and the position of the interconnector through the partition wall of the container. The feet on which the plate stands on the upstanding ribs in the base of the container are positioned according to the position of those ribs. The purpose of the container ribs is to create a 'sump' or sludge space into which active material which has become detached from the plate surface can collect without harming the function of the battery. The feet of opposing plates rest on alternate container ribs, to prevent the possibility of material trapped on the top of the container rib from causing a short-circuit. In cells with enveloped plates such traps are not needed as the material from the plates of one polarity is contained within the plate envelopes. The precise sections, angles and radii of the parts of the grid are a function of the casting alloy used, its fluidity and freezing characteristics, the moulding tool technology used and the casting conditions that apply.

The pitching of the strands and ribs defines the 'pellet' size, the rectangular volume of paste surrounded on two sides by ribs and on two sides by strands. Too large a pellet size leads to inefficient use of active material as the part of the pellet at greatest distance from a conductive path contributes less to the charge-discharge reactions. The inherent strength of the plate is also somewhat reduced. Too small a pellet size and capacity is sacrificed as a larger percentage of the total plate volume is taken up by inactive grid material.

As discussed elsewhere the active material consists of a porous matrix of small irregularly-shaped particles with a void space of about 70%. The electrons produced by, or required by, the reactions taking place within the matrix must follow conductive paths from particle to particle and travel readily to and fro across the grid/active material interface.

In the case of the negative plate this is a comparatively simple task, as the particles in the charged state are lead, and even in the discharged condition not all of the lead is converted into lead sulphate (as noted elsewhere, active material utilisation is of the order of 50% even at very low rates of discharge). Hence a considerable matrix of conducting lead is always present and the rib and strand pitch of the negative grid can be much wider. The negative grid is also not subject to electro-chemical corrosion in the same way as is the positive grid, and the grid cross-sections can therefore be much lighter.

In the case of the positive plate, however, the active material consists of particles of lead dioxide in the charged state. Lead dioxide is a non-stoichiometric compound. In other words there are not exactly two oxygen atoms for every lead atom. There are rather less than two oxygen atoms per lead atom throughout the crystal lattice-work, and this gives lead dioxide semi-conducting properties. In addition, during charge the grid of the positive plate can be subjected to corrosion, adding to the layer of lead dioxide covering its surface. Hence voltage-drops across the positive plate are higher than those across the negative plate, and grid sections tend to be thicker to compensate for corrosion. The design of the grid clearly influences the way in which current, generated by the plate active material, flows along the conducting paths of the grid, concentrating towards its exit or entry at the lug. The current flows can be shown to largely follow Kirchhoff's Laws, and potential contour lines can be drawn for a plate of a particular design (Puzey and Orriel, 2.18.). Attempts are made to increase conduction towards the lug by increasing the density of stranding, increasing sections of ribs and strands in these areas or adding ribs angled down from the vicinity of the lug, often referred to as a radial design. (See Figure 2.17.)

Ideally the lug should be in the centre of the plate but this clearly is not practical. Designers attempt to put the lug as near the centre of the top frame as possible in order to minimise voltage drop on the positive plate. (This position also provides increased resistance to vibration.)

The use of lead/calcium alloy expanded-metal grids has increased greatly in recent years, mainly for negative plates, but as mentioned earlier increasingly in positive plates. Lead/calcium alloys reduce the amount of water-loss and as an additional benefit are more conductive than conventional lead/antimony alloys. A typical negative grid is shown in Figure 2.18. Positive grids have a smaller diamond lattice. In general the diamond has a horizontal to vertical chord ratio of about 1.5/1. A lower ratio than this would subject the metal to an undue amount of stress at the intersections or 'nodes', leading to cracking. The expanding action provides some twist in the wires between the nodes, leading to further stress but providing good angles for paste to 'key' onto it.

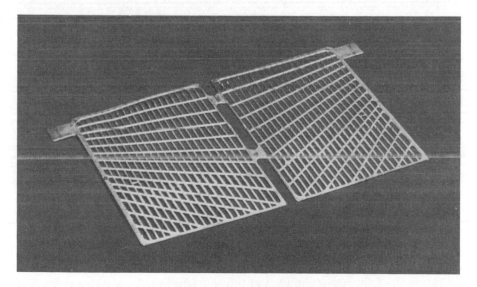

Fig.2.17. Photograph of positive grid with radial design.
(Lucas Yuasa Batteries Ltd)

Expanded-metal negative plates are often associated with cast low-antimony positive plates in a so-called 'hybrid' construction. Early batteries containing positive plates with lead/calcium alloy grids suffered from problems associated with recovery of charge from the discharged state. This was found to be due to the formation of an inert boundary layer between the grid and the active material. While the newer alloys containing calcium and tin (at higher levels than previously used) appear to have overcome this problem and operate very successfully in automotive applications, there remain some questions regarding the capability of such batteries in deep-cycle applications.

Cell capacity is a function of the total mass of the positive and negative material available, together with the amount of sulphuric acid present. In theory, cell capacity could be achieved by having one thick plate of each polarity in a bath of sulphuric acid. Such an arrangement would be very inefficient for a number of reasons.

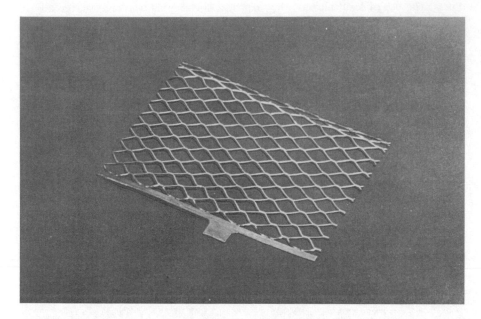

Fig.2.18. An expanded metal negative grid
(Lucas Yuasa Batteries Ltd)

Most of the cell reaction takes place on adjacent surfaces. (The ionic stream through the electrolyte follows the shortest path across the potential difference.) The deeper into the pores the plate reaction takes place, the quicker the performance falls off due to the finite rate of diffusion of sulphate ions in and out of the pores. The higher the current required, the faster the reactions have to be driven and the faster the performance dies away.

In practice surface area should be made as large as possible to achieve good material-utilisation. The real surface area, i.e., that of all the particles less the touching areas, is clearly many times greater than the apparent surface area, that defined within the perimeter of the grid frame. Real surface area is maximised by using particles of as irregular a shape as possible with a spread of particle size. Gas absorption techniques are normally used to gain a measure of real surface area, but double-layer capacitance measurement has also been used (Amlie, Ockerman and Ruetschi, 2.19). Mathematical models describing the electrochemical behaviour of porous electrodes have been developed (Newman and Tiedemann, 2.20.). However, suffice it to say here that too open a structure gives rise to greater fragility, and more material shedding with cyclic use, thus limiting life. Too closed a structure reduces material utilisation. Once a practical compromise has been reached, thus defining the paste formulation, the designer, in practice, uses the apparent surface area for his calculations. Hence many thin plates give rise to better utilisation at higher currents. However there are limits in this direction as well. As mentioned elsewhere there are practical limitations on as-cast section thicknesses. Very thin plates become much more difficult to handle in production. In practice the proportion of grid material to active material starts to rise again, leading to a reduction in power/weight ratio and power/volume ratio.

The other component of the cell pack is the separator. The prime function of the separator is to prevent electronic contact of any sort between the positive and negative plates of a cell, but at the same time to create minimum resistance to the passage of ions from plate to plate. The closer the plate-to-plate spacing the higher the voltage obtained at any given discharge current, but the higher the risk of crystal growth resulting from electrochemical action, forming a plate-to-plate electronic bridge, known as 'treeing'. Clearly, from the above, it is important to have separators that are as thin as possible, but as noted earlier, it is also important to have adequate acid available to the plates. This is particularly so at the positive plate, and it is normal to design the separator with vertical upstanding ribs to allow for this. However, the total separator area occupied by such ribs should be kept to a minimum to ensure that ionic conductivity remains high. In addition a smooth surface in contact with a plate can lead to entrapped gas-bubbles which can prevent large areas of the paste from functioning, so small ribs are also put on the side of the separator exposed to the negative plate to avoid this situation. Figure 2.19 shows a section through a typical separator.

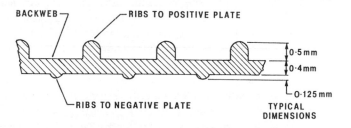

Fig.2.19. Section of a typical separator

One area that over the years has occupied much development testing with no clear-cut result is the question of the outside plates of the cell. As pointed out previously the majority of the reactions take place on adjacent surfaces. Hence the surface in proximity to the cell wall plays only a minor role, particularly at high rates of discharge. Cell walls are normally ribbed to aid acid availability at these surfaces, which enables some low-rate reaction to take place. Traditional battery design assumed an odd-number of plates, with the outer plates being negatives. Excess negative material was always built-in on the basis that at low temperatures high rate discharge was limited by the negative material. Special thin outer plates were sometimes manufactured. Some manufacturers used outer positive plates. In recent years there has been a trend towards an even number of plates. In any event, as a greater number of plates per cell has become the norm, so the difficulty reduces in importance.

The acid used in the battery as an electrolyte is a dilute form, with a specific gravity of 1.26-1.28 (15°C), or a concentration of 34-37% by weight of sulphuric acid in water.

The choice of this concentration of acid, adopted almost universally for temperate climates (acid of somewhat lower concentration is often specified for tropical use), is a compromise between a number of factors, recognising that the acid takes part in the cell reactions and is not just present as an ionic conductor, as is the electrolyte in most other electrochemical cells.

1) The solution becomes depleted of sulphate ions during discharge. Hence the total sulphate ion concentration in the fully charged battery must largely balance the available reaction materials of the plates.

2) Conductivity is at an optimum between specific gravities of about 1.1 and 1.3.

3) Unwanted corrosion and self-discharge reactions increase with increasing acid concentration.

4) Viscosity increases with increasing acid concentration (and decreasing temperature).

5) The working temperature range of a typical automotive battery is from -20°C to 70°C (although industry specifications call for wider temperature limits to cater for extreme conditions). At the lower end of the temperature scale electrolyte can freeze if the concentration is too low. At the upper end of the range other unwanted chemical reactions and self-discharge start to become important factors.

The connector strap with its interconnector extension is a key component demanding close attention. The designer is always under pressure to make the connection of the grid to the strap as close to the top of the grid as possible and to make the strap and interconnector as small as possible, firstly to minimise the electrical path length and secondly to save weight and therefore cost. However, he must be mindful to make sure that the design is capable of easy manufacture and assembly, that the materials of the grid and the strap are compatible to ensure that good fusion takes place, to minimise ohmic losses, and prevent the possibility of plates becoming disconnected through vibration, shock or the effects of electrochemical corrosion. However, thinning or melting of the grid lug below the strap while joining takes place must be prevented. In addition, close proximity of the connector-bar with the top of the plates of opposite polarity can give rise to shorting, due to the build-up of material on the top edges of the plates. The interconnector extension must be designed to provide an adequate sealing-land around the partition hole, to ensure that only sufficient lead to fill the partition hole on fusion during production is provided, thus preventing any splashing out of the briefly-molten lead. Too little material and the weld is of inadequate strength and cross-section. The connector and interconnector unit must be designed in a manner that prevents undue distortion of the plate group within the cell. Finally, of course, the assembly must be capable of carrying the maximum current, which occurs when the fully charged battery is short-circuited, without fusion of any element of the assembly.

Most modern batteries are designed so that the cell pack is a reasonably tight fit within the side ribs of the cell partition walls to resist vibration damage. This particularly applies to batteries containing enveloped plates. These tend to have very thin fragile plates and the structural strength is more dependent on the binding of the pack within the cell compartment than the older designs where individual component strength and rigidity are vital. In such batteries, where they are exposed to very severe vibration or shock environments, the cell design will sometimes incorporate additional features to combat these circumstances. The plate-foot design can be modified so as to lock the plates onto the footrest. Plastic devices bridging the top of the plate group can be incorporated. Ribbons of cementing material across the bottom and top of the plate group are also used.

2.8 External Design Details

Most battery users are aware of the taper-post terminal used in the majority of installations globally for many years. It offers a good, reliable area of contact with the wire termination from the starter motor, and is easy to maintain. However it is bulky, as is the wire termination, not easy to protect, and the positive and negative terminals are not easy to differentiate visually or by the wire terminations. Much of the electronics used on a modern vehicle is susceptible to damage if the voltage is connected the wrong way round. Hence it is important to find means to minimise this danger. Vehicle manufacturers tend to reduce starter cable lengths to a minimum to save voltage drop, weight and cost. This and the fact that most batteries have the terminals to one side or the other tends to minimise the problem. If the correct battery is fitted the wrong way round, the cable terminations in fact will not reach. It does nothing, however, for the situation where the wrong polarity battery is offered up.

Over the years many attempts have been made to overcome this and other termination problems. One large vehicle manufacturer has used nut and bolt spade terminals for a number of years, and by clever design incorporating some plastic insulating parts, a degree of 'go/no go' capability has been achieved. Side-wall terminations have been used by another large manufacturer with similar capability (Figure 2.20). Double nut-and-bolt terminations are used by some manufacturers of large vehicles where very large starting currents are required. (See Figure 2.21.) Flying-leads and plugs and sockets are often used for motorcycle batteries. However, whatever the form of battery termination designed or specified, the battery designer must ensure that the end-cell to terminal connection is properly sized for the maximum current condition, that the joint between the cover or container and the connector is leakproof, and that the strength and torque-resistance of the terminal meet the specified requirements. He must consider protection of the terminal, perhaps by

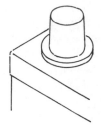
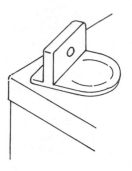
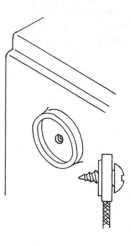

Fig.2.20. Schematic showing various terminal designs

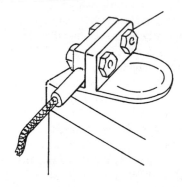

Fig.2.21. Double nut-and-bolt terminal

recessing the terminal or by the use of protective covers, and is often called on to design the cable termination as well, to ensure compatibility The manner in which the battery is clamped in the vehicle needs to be defined at an early stage. The traditional top-frame or top-edge clamps with 'J' bolt fastening is expensive and not simple to fit. There is also the possibility of shorting during fitment or resulting from corrosion. Cross-battery straps restrict access to the vent apertures. Clamping by means of ledges moulded into the sides or ends of the container at the base offers an inexpensive solution to most of the above problems and has been adopted by many manufacturers.

The design of the battery cover to prevent gas-entrapment, to incorporate vent plugs, filling manifolds, pressure relief-valves, nozzles for piping away gases and explosion-proof vents are all areas needing to be defined.

2.9 Production Aspects of Design

It has been mentioned that the development of thin plates has enabled the battery designer to 'tailor-make' batteries for specific requirements. This statement needs to be tempered with the production need to keep components and assembly as standardised as possible.

Modern automated production-machinery requiring high capital investment must be justified on the grounds of high volumes of production with minimum downtime for changeovers or adjustment.

The production team is always under pressure to keep work-in-progress stocks to a minimum (manufacturing processes are such that in most cases a minimum of 5 to 7 days is required to convert lead and lead alloy into a finished battery) while at the same time being able to respond flexibly and quickly to customer demands.

Hence, particularly in the plate-making stages it is important to keep design variants to the minimum. (It is also at times quite difficult to differentiate plate stocks where weight and thickness specification differences become quite small.)

The capital cost of tooling up for a completely new design is very high, in particular the moulding tools for container and lid, which may amount to many tens of thousands of pounds.

It goes without saying that the designer must be aware of the capability and limitations of the plant and equipment at his disposal. It is vitally important to minimise scrap and downtime. Handleability can become a major issue as components become lighter and more fragile, and are processed at ever increasing speeds.

2.10 Standards and Standard Design Specifications

There are a number of different standards governing the design of the automobile battery, with most of them covering the same areas of performance criteria and test methods.

There is an international standard covering electrical performance generated by the International Electrotechnical Commission (the IEC) which is supported by the Standards bodies of a number of countries and whose individual standards follow the IEC standard to some extent. These include most of the countries of Western Europe.

As with most international agreements, however, the IEC standard, No. 95, Parts 1, 2, 3 & 4, due to bureaucracy and the democratic nature of its generation, tends to be more generalised and difficult to maintain up-to-date, compared with the more specific standards of the individual countries. Thus, while it is useful to have some internationally recognised benchmarks, the IEC standard is generally recognised as the minimum requirement. National standards usually cover dimensional and other physical requirements in addition to electrical performance. In Europe, due to the strength of the German automotive industry the DIN (Deutsche Industriale Normale) standard, No. 43539, is generally recognised as being the most useful and appropriate. The USA looks to the Society of Automotive Engineers, the SAE, to generate and maintain its automotive battery standard, No. J537, while the other key vehicle-manufacturing country, Japan, uses the Japanese Industrial Standards body, the JIS, No. D5301.

Each one differs in detail, but in general they all set down test methods, standard environmental conditions and a sequence of testing new batteries in an attempt to provide comparative data from accelerated laboratory testing which simulate as closely as possible real life conditions. They cover:

a) the engine starting situation by measuring the current that the fully charged battery can deliver at a realistically low temperature (normally about -18°C) to a working minimum voltage in a given time, e.g., SAE defines 30secs. to 8.4V, IEC 10secs to 7.5V and JIS 30secs to 7.2V.

b) the low rate capacity required. Traditionally the ampere hours available from the battery when discharged at a very low current, e.g., one sustainable for 20h at room temperature (normally defined as 25°C), was used. In recent years the much more realistic situation of coping with the vehicle load under non-charging situations has been used. A current of 25amps has been defined as the load current, and the time taken, at 25°C, to reach a predetermined voltage, often 10.5V, used as a measure of the battery's capability.

c) charge acceptance. This measures the ability of the battery in a partially-charged condition and at a low temperature, normally 0°C, to be recharged at the level of constant-voltage available from a vehicle generator, about 14.2V for a 12V battery.

d) charge retention. This test defines the ability of a battery to hold charge by requiring it to provide a measure of starting current at low temperature after prolonged storage at high temperature, (the IEC specifies 40°C).

e) life or endurance testing. This is the test most fraught with difficulty in attempting to simulate real life conditions, and gives rise to the most variation in the methods adopted by the various standards. In concept the test tries to stress the battery by a series of partial charge/discharge cycles under high/low temperature conditions, in a manner that would give rise to eventual battery failure.

f) water retention. This is a fairly new test aimed to cater for the new generation of maintenance-free batteries. Water mass loss is measured after a standard level of overcharge is applied, normally at a high temperature.

Most of the large vehicle manufacturers develop their own standard requirements, which are the most detailed and onerous, covering very specific design details, performance requirements, physical requirements, often material requirements, test methods and certification requirements. It is against these, sometimes confidential, documents that the battery manufacturer designs, develops and supplies batteries to the vehicle manufacturer.

However, in developing these standards the vehicle manufacturer, for reasons of economic production and availability of spares, replacement, etc., normally takes cognisance of current standard-practice and state-of-the-art. Furthermore, as vehicle manufacturing becomes more global in its approach it is likely that the number of standards will become fewer and feature more common requirements.

One more group of standards should be mentioned, namely the military standards. Batteries for military vehicles, and in particular the front-line vehicles known as fighting vehicles, including tanks, gun carriers etc., tend to be of specialised design and performance to cater for the more extreme and rigorous environment that these vehicles experience. Vibration and shock levels tend to be much higher than those in more conventional vehicles, and temperature requirements as low as -40°C are often specified.

However, even in this area, due to the relatively small volumes involved and the high costs of production, military vehicle designers are looking more and more towards solutions using commercially available batteries.

2.11 Application of Automotive Batteries in other Fields

Due to its mass availability and relatively low cost the automotive battery finds application in a large variety of fields for which it was not intended or designed. (For this reason most manufacturers specifically exclude such applications in their warranty cover.) However, in many of these applications it has been found to be the most practical and appropriate solution.

Applications can be divided into those associated with vehicles, traction use, and other applications, but all these, in general, use the low-to-medium rate capability and operate in a cycling mode, as distinct from the high-rate discharge but otherwise float-mode of the automotive application.

Those associated with vehicles in general use the vehicle charging system to replace lost capacity. Automotive batteries are often found in trailers or caravans where they are used for domestic purposes, lighting, powering small pumps, etc. The battery is used at low rates, about C_{10}, and is frequently deep-discharged. (See Section 5.2 for an explanation of the C_x rate.) A much heavier-rate application, C_5, is where a separate battery is used to power tailgate lifts on trucks, or pumps on tankers. These are not normally deep-discharged.

Traction applications, requiring current at around the C_2 to C_5 rate, include electric wheelchairs and electric cycles. These are often deep-discharge uses requiring rapid recharge, and are among the most onerous and life-limiting applications. What may be classified as semi-traction applications include lawnmowers and golf trolleys, again perhaps the C_5 rate, but with intermittent, often seasonal use. (In these applications battery life is usually restricted by abuse caused by leaving the battery in a discharged state for long periods.)

Other applications, particularly in the developing world where mains electricity has yet to be made universally available, include domestic lighting, televisions etc., requiring C_2 to C_5 currents. Sometimes batteries are connected to solar panels for recharge, but often a vehicle is used

Where the market is of sufficient size, appropriate changes to the basic automotive design can be incorporated. These include the use of heavier grids, perhaps fewer in number, with more dense active-material and thicker, more resilient separators. Such batteries are available in some areas for golf trolley and electric wheelchair applications, for example.

One further very important sector is the starting of stationary engines used for pumps, generators, compressors etc. Weight and size are not normally important considerations, and the battery is often not subject to severe vibration or large temperature excursions. On the assumption that charge regulation is good, this can be considered to be a light-duty application.

2.12 Valve Regulated Batteries

On overcharge, batteries containing aqueous electrolytes, with lead-acid and nickel-cadmium being the prime examples, evolve hydrogen and oxygen by simple electrolysis of the water present.

In a conventional battery this is allowed for by enabling the battery to vent these gases to atmosphere. Most applications, including automotive, employ charging systems that sense when the battery is becoming fully charged, and reduce the charging current accordingly, thus limiting the degree of overcharge and hence water loss. However, some water loss does take place, and water needs to be replaced at intervals. In addition the battery can leak electrolyte, for example if tilted or inverted, or if overfilled. (Sulphuric acid has a very low surface tension and is notorious for creeping over surfaces, wetting them and forming conductive paths. Hence even the external lid surfaces of automotive batteries are often wet with acid even when none has been spilled.)

It is theoretically possible to recombine the gases catalytically to water, and therefore seal the battery. In practice the gases are rarely evolved stoichiometrically, and there is therefore always an excess of one of the gases, and it is difficult to devise a practical, reliable and cheap catalyst system. (See Section 2.5.1.)

However, it has been shown that under the right circumstances gases, and in particular oxygen, can be made to recombine at the opposite electrode to that at which they were evolved. In essence the right circumstances are where a three-phase interface exists on the recombining electrode, i. e., where the electrode surface, the gas and the electrolyte co-exist. The actual reactions taking place are much more complex than indicated here, but equation 2.21 summarises the oxygen evolution process taking place on the positive electrode while equation 2.22 shows the absorption process on the negative electrode.

$$2H_2O = O_2 + 4H^+ + 4e^-$$

$$(2.21)$$

$$2Pb + O_2 + 2H_2SO_4 = 2PbSO_4 + 2H_2O \qquad (2.22)$$

A number of manufacturers have reached this situation in practice by developing systems that use an immobilised electrolyte. This is achieved either by using very finely-powdered silica or alumina to create a 'gelled' electrolyte or using a mat of very fine glass fibres which wholly absorb the electrolyte. In either case free mobile electrolyte is unavailable to fill all the space next to the plate, thus providing the opportunity for gas to reach the plate surface.

Batteries employing this technology have been shown to have a very good level of gas recombination, encouraging manufacturers to produce batteries that to all intents and purposes are sealed. They all employ safety or regulator valves to prevent excessive build-up of gas within the unit, but can be used in any attitude and are leakproof.

Such batteries have found application where current loads are moderate, charge control is good and temperature variations are not excessive, e.g., Uninterruptible Power Supply (UPS) systems, portable electronic equipment and, more recently, cordless tools.

Application in the automotive field has, to date, been very limited, due mainly to two key factors. The first is the inability of the battery to deliver and sustain quite such high levels of discharge current as is available from the standard automotive battery, and the second is the necessity for very sophisticated charging control, and a consquent vulnerability to high under-bonnet temperatures.

However, these difficulties are being addressed, and there is little doubt that in years to come valve-regulated batteries will have widespread use in the automotive field.

REFERENCES

2.1 'Physical Chemistry' (5th Edition) by P.W.Atkins, Oxford University Press,1994.

2.2 'Physical Chemistry' by G.M.Borrow, 6th Edition, McGraw Hill,1996.

2.3 'Reference Electrodes' by D.J.G.Ives and G.J.Janz, Academic Press (New York),1961.

2.4 'Electrode Kinetics' by J.Albery, Clarendon Press, 1975.

2.5 'Electrochemical Techniques: Fundamentals and Applications' by A.J.Bard and L.R.Faulkner, Wiley-Interscience, New York, 1979.

2.6 'Electrochemical Kinetics' by K.J.Vetter, Academic Press, New York,1967.

2.7 W.Peukert, Elektro-Technische Zeitung, 18 (1897), 287-288.

2.8 C.A.Fauré, Compte Rendu, 92(1881), 951-953. Also French Patent No. 139,258(1880).

2.9 'Lead-Acid Batteries' by H.Bode, Wiley, New York,1977.

62

2.10 'Current Theory of Lead-Acid Batteries' by M.A.Dasoyan and I.A.Aguf, (English Translation), Technology Ltd. (In association with the International Lead and Zinc Research Organisation Inc.), 1979.

2.11 'Research in Lead-Acid Batteries' by J.Burbank, A.C.Simon and E.Willinghanz in 'Advances in Electrochemistry and Electrochemical Engineering', Vol. 8, Ed. C.W.Tobias and P.Delahay, Wiley, 1976.

2.12 G.Hoffman and W.Vielstich, J. Electroanalytical Chem., 180, 565, 1984.

2.13 'Battery Expanders and their Use', by G.Szara in 'Improvements in Alloys, Oxides and Expanders for Lead Batteries', Lead Development Association, London, 1984.

2.14. G.R.Webley, Proc. 2nd International Symposium on Batteries, Bournemouth, 1960 (Joint Services Power Sources Committee).

2.15 'Charge Acceptance of the Lead-Acid Cell' by K.Peters, A.I.Harrison and W.H.Durant in 'Power Sources 2', Ed. D.H.Collins, Pergamon Press, 1968.

2.16 'Gas Recombination Lead-Acid Stationary Batteries' by A.I.Harrison and B.A.Wittey in 'Power Sources 10' Ed. L.Pearce, Paul Press, London, 1985.

2.17 'Sealed Lead-Acid Batteries for Aircraft Application' by J.Thompson and S.Warrel, in 'Power Sources 9' Ed. D.H.Collins, Pergamon Press, London, 1982.

2.18 'Distribution of Potential over Discharging Lead-Acid Battery Plates', by J.E.Puzey and W.M. Orriel in 'Power Sources', Ed. D.H.Collins, Pergamon Press, 1970.

2.19 L.R.F.Amlie, J.G.Ockerman and P.Ruetschi, J. Electrochemical Soc. 108 (1961), 377-383.

2.20 'Mathematical Modelling of Phenomena contributing to Thermal Rise in Lead-Acid Batteries used in Electric Vehicles' by W.Tiedemann and J.Newman in 'Advances in Lead-Acid Batteries', Ed. K.R.Bullock, D.Pavlov, Electrochem Soc. Proceedings, Vol 84-14, Electrochem Soc. Pennington N.J., 1984.

CHAPTER 3

Lead-Acid Battery Manufacture

Lead-acid battery manufacturing is one of the few parts of the engineering industry that combine chemical processing, both continuous and batch, with mechanical processing. The key material, lead, is both heavy and, in metallic form, soft. It is also toxic to a degree. The process uses sulphuric acid which is highly corrosive. Added to all this, the product in wet-charged form is perishable! The industry has, in years gone by, gained a reputation for being at the dirty, low technology, end of manufacturing. This position has changed dramatically in the last twenty years or so, influenced by the development of new materials, in particular plastics, the demand for more maintenance-free products, the development of modern automated equipment and the demand for ever higher standards of consistency and reliability from the product.

An overview of the manufacturing process can be seen in the flow chart shown in Figure 3.1.

3.1 Lead Oxide Production
Early batteries were made using pure lead monoxide powder, (litharge), but it was soon discovered that battery plates made with incompletely oxidised lead powders provided far superior cohesion of the particles within the paste, and hence greater battery life. Precise specification of these 'leady' lead oxides varies from manufacturer to manufacturer but this material is in universal use today, normally containing about 25-40% of unoxidised lead.

There are two main production techniques in use, the Ball Mill process, shown schematically in Figure 3.2, and the Barton Pot process, shown schematically in Figure 3.3. Each process has its own dedicated followers, but although there are significant differences in particle size, particle shape and size distribution, all of which are important parameters for subsequent processing, there would appear to be little difference, either in terms of ease of production or ultimate performance, between the two. However, the processing differences are such that mixing the two oxides is not practicable.

Whichever process is chosen, perhaps the most important consideration is that,

64

having established subsequent processing parameters, the oxide should be of as consistent a quality as possible.

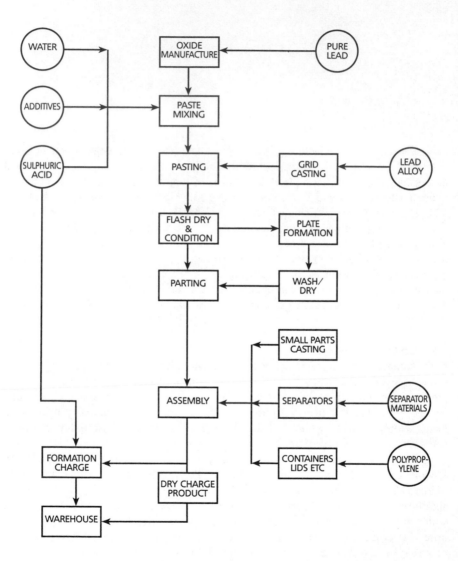

Fig.3.1. Flow chart showing the automotive battery manufacturing route

3.1.1 The Ball Mill Process

Pure lead in the form of balls, chopped bar or even whole ingots, depending on the manufacturer's process, are automatically fed into a large steel drum, typically of about 2m diameter, electrically driven, rotating on a horizontal axis and through which air is blown. The rotational speed is such that the material in the drum cascades over itself and does not centrifuge. Heat is generated by the friction of the material continually tumbling in the mill and the oxidation itself is exothermic. The grinding action produces fine particles of partially oxidised lead, forming a powder which is taken away in the air stream and collected in a cyclone from which it is either drummed or conveyed to storage silos. Ultrafine particles are collected in a subsequent fabric filter and added back into the main fraction.

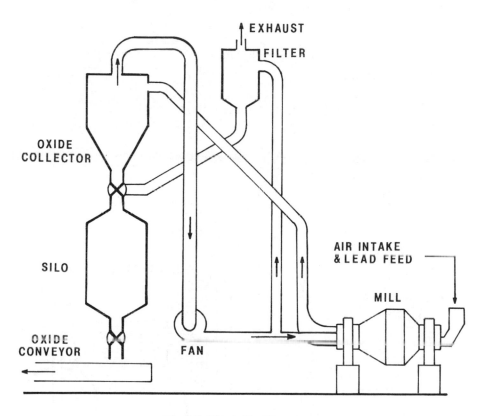

Fig.3.2. The ball mill process

Oxide output and specification can be affected by a number of variables, including mill speed, amount of load, temperature of load, air velocity and air temperature. In modern ball mills, all these parameters are monitored by sensors

66

which enable automatic adjustments to be made on a continuous basis, thus maintaining an equilibrium condition and a consistent product. Such an installation would be expected to run continuously with no more than a machine minder, 24 hours a day, normally with a shutdown for cleaning, maintenance etc. on a weekly basis.

3.1.2 The Barton Pot Process

The Barton Pot process operates on an entirely different principle. Molten lead from a holding pot is pumped at a controlled rate into a reaction chamber and agitated strongly by a rotating paddle. Air is drawn through the reaction chamber at a controlled rate, which oxidises small droplets thrown up, and carries the particles of lead oxide so produced to a cyclone, from where the material can be drummed, or conveyed to silos. Sometimes the oxide is subjected to a hammer mill to flatten and increase the surface area of the particles.

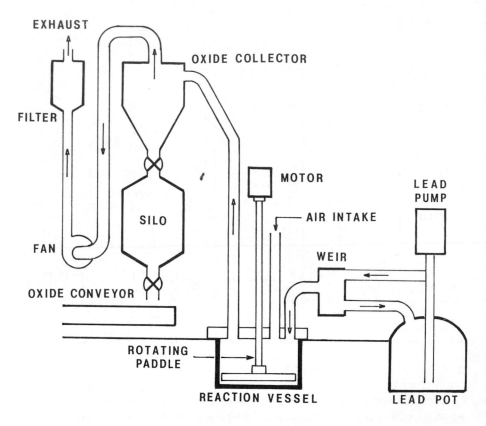

Fig.3.3. The Barton Pot process

Oxide output and specification are dependent on the flow rate and temperature of the molten lead, air flow rate, paddle wheel speed etc., and again can be controlled automatically.

The Barton Pot has the advantage that it can be stopped and started again more conveniently than the Ball Mill, but this is offset by greater difficulty in holding equilibrium conditions.

With both processes it is essential to start with pure lead, 99.99% being the normal minimum purity level specified, although in the case of the Barton Pot process traces of antimony and copper have been claimed to aid oxidation. Whereas most manufacturers now use secondary lead for grid and connector purposes, many still rely on primary refined lead for oxide production, on the basis that it contains less trace element impurities, and therefore gives longer battery life.

As mentioned above, the oxide produced by the two processes is significantly different. Ball Mill oxide consists of irregularly shaped platelets with a mean particle size of perhaps 3-4μ (1μ=1/1000mm) but a fairly wide distribution including a significant proportion of finer particles and an unoxidised lead content of 25-30%, whereas Barton Pot oxide has a more spherically shaped particle of larger average size, 7-8μ, with a close distribution spread and a lead content of up to 40%.

In addition to the automatic machine controls it is usual to check regularly the oxide at 2-4 hr intervals, measuring the metallic lead content and the 'Scott cube' or apparent density (the weight of a lightly compacted known volume of material).

All parts of the mechanical system for capturing, storing and taking off oxide need to be well sealed to prevent escape of material into the atmosphere and possible hazard to the operators working in the area.

3.1.3 Storage of Oxide

Oxide is normally stored either in sealed steel drums, conveniently sized for subsequent batch paste mixing, and typically containing about 200 kg, or in large silos of up to 100 tonnes capacity, from which oxide is drawn off on demand. However, as the powder particles contain a core of metallic lead, which is reactive in moist air, generating heat, care needs to be taken, particularly when storing in bulk, to minimise oxidation and avoid the possibility of thermal runaway taking place. Bulk silos are normally fitted with a system which causes them to be flooded with gaseous nitrogen if an abnormally high temperature is sensed.

3.2 Grid Production

The grid has two main functions: to provide a support for the active materials taking part in the reactions, and to conduct the current resulting from those reactions. In addition it must be able to withstand the corrosive effects of the electrolyte, not interfere with the reactions required, and not introduce unwanted electrochemical reactions. Thus, despite being only a moderately good conductor, heavy and relatively soft, the natural and indeed universal choice of material for the grid is lead. In recent years lead or titanium coatings onto rigid plastic substrates, and carbon have been tried but to date no successful substitute for alloyed lead has been found.

Pure lead, while being chemically and electrochemically the best material is too soft, particularly for handling during processing. It is therefore alloyed to provide some stiffness and for casting to aid fluidity. The alloying of lead to provide the best set of characteristics for use as a grid material has been the subject of a great deal of research, and many patents have been filed. Suffice it to say here that traditionally, for gravity casting, antimony has been the favoured element in the range of 1.5-10%, often associated with small amounts of tin and arsenic, the former to aid fluidity and the latter for stiffness and tensile strength, particularly at lower antimony levels. At the higher levels of antimony the alloy approaches the eutectic at 11.1%, melting at $252^{o}C$, and castability is good. However, antimony is undesirable electrochemically as it dissolves from the positive plate, migrates to the negative plate, causing self-discharge and a lowering of the hydrogen overpotential. Low antimony alloys prove more difficult to cast however, requiring greater control of alloy content and casting conditions. Grain refiners such as sulphur or selenium are often included. Newer alloys, without antimony, have been developed enabling batteries to become more maintenance-free. Binary alloys containing small amounts of calcium, of the order of 0.1%, are used for negative grids, with tin being added as well for positive grids, at levels of approximately 0.2%. Strontium has been used instead of calcium in some alloys, and lead/calcium/aluminium alloys have also been used. In these last mentioned alloys the aluminium is present to protect against loss of calcium in the molten alloy. The mechanical characteristics of the lead-calcium alloys make them very suitable for expanded metal production. Antimonial alloys have been prone to cracking and creep problems in expanded metal form, but it is anticipated that manufacturing technology and further alloy development will allow for low antimonial alloy expanded grids in the near future.

In recent years as vehicle systems have improved, the role of the battery has changed, with a greater emphasis on high-rate currents for starting. This is achieved by packing the maximum number of thin plates into the cell to provide the maximum surface area. Hence modern technology has concentrated on the manufacture and handling of thinner and thinner grids.

There are currently three methods in general use for manufacturing grids.

3.2.1 Gravity Casting

This is the traditional, and still the most widespread method today, using automatic machines similar to that shown in Figure 3.4.

Molten alloy is held, in either a gas- or electrically-heated pot, fed manually or automatically by ingot, to maintain a consistent level, at about 350-450°C depending on the alloy, and pumped or gravity fed, in a metered amount, to the top or gate of a pair of vertically supported half moulds similar to those shown in Figure 3.5. The alloy is superheated by about a further 50°C on the way to the mould, into the abutting faces of which are cut the grid form required, normally as an opposed double pair to aid subsequent handling (see Figure 3.6). The molten metal is then allowed to flow down under gravity to fill the mould. When the alloy has solidified, the half moulds open, and the double grid is ejected to fall onto a moving belt, and thence onto take-off arms.

Fig.3.4. A modern automated grid casting machine
(Wirtz Manufacturing Co. Inc.)

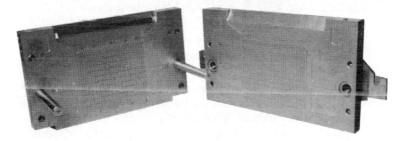

Fig.3.5. A typical grid casting mould
(Wirtz Manufacturing Co. Inc.)

The moulds are normally electrically heated to about 150°C and water-cooled. The abutting faces are dressed with a cork compound, providing insulation and some

70

air venting. The design and control of such moulds, to provide fast and efficient filling with adequate air venting while avoiding defects in the product, has been one of the more important improvements of recent years. Poorly controlled filling can give rise to incomplete filling and air entrapment thinning in the complex cross stranding of the grid. Hot cracking or tearing can occur if freezing is incomplete on ejection, often not apparent to the naked eye. All of these defects can seriously increase the corrosion of the grid and therefore shorten the battery life. Flashing and blinding, (that is, the situation where the spaces between the ribs of the grid become filled with a thin web of metal), can occur if the metal is too fluid, venting too free, or there is a lack of shutout between the two halves of the mould; and these defects cause poor paste adhesion and shorten battery life.

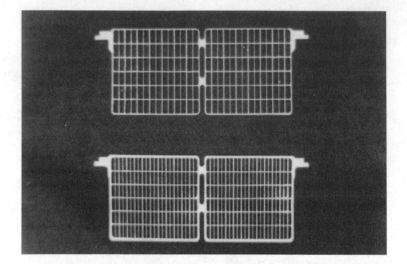

Fig.3.6. As-cast positive and negative double grids
(Lucas Yuasa Batteries Ltd)

Automatic casting machines with the latest-technology grid moulds typically cycle at 16 to 18 castings a minute using either 1.5-2.5% antimonial lead alloy or about 0.1% calcium lead binary or tertiary alloy (including small additions of tin) with grid thicknesses down to about 0.75mm, considered by many to be the limit to which the conventional gravity casting process can be taken. The latest automatic machines have a microprocessor control system. Temperature is sensed and controlled in the holding pot, feedline, ladle and at various points on the mould. Speed of casting is automatically controlled. Problem diagnostic messages are shown to the operator on a screen.

The mould dressing (normally cork), needs repair or replacement usually once or twice per eight hour shift, and the state of the dressing tends to be one of the main preoccupations of the machine minder, who will be constantly vigilant for casting defects. The machine minder will typically be looking after four such casting machines and his other roles include regular checks of thickness, weight and absence of distortion, ensuring that pot levels are maintained, some de-drossing from the surface of the pots, and emptying the take-off arms of the accumulated finished double grids, stacking them onto wooden pallets for storage. Regular analysis of pot metal should be carried out, as some losses of alloying metal or other element (e.g., sulphur is often incorporated as a grain refiner) can occur through oxidation or evaporation. Sometimes an inert atmosphere is used above the molten metal to minimise oxidation, particularly when casting alloys containing calcium. Often a minimum storage period of 24-48 hours is a requirement for some age hardening of the grid alloy to take place before subsequent processing. For some alloys an immediate water quench after casting has been found to be beneficial, as this stiffens the grids and aids handling.

Molten lead pots are normally well shielded to prevent splashing, and hooded to facilitate fume extraction. Other safety precautions are similar to those found in any foundry environment.

Ingot stocks must be kept dry to prevent explosive evaporation of residual or entrapped moisture when fed to the pot.

Care must be taken when using both lead-antimony alloys and lead-calcium alloys in the foundry, as mixed molten metal can give rise to drosses containing calcium arsenide and calcium antimonide. When wetted, these materials react with water to generate poisonous arsine and stibine gases. Similar problems can occur at the secondary smelter when alloys are inadvertently mixed.

3.2.2 Expanded Metal Grids

The fabrication of grids from continuous strips of alloy offers the opportunity for considerably higher production, and productivity rates, than those achievable with gravity casting.

Two techniques, first developed in the 1970s, are in widespread use today. The first of these uses wrought lead/calcium alloy strip. This is rolled to a given thickness (0,6-0.8mm), slit to a predetermined width, normally about 75 mm for a typical car battery grid, and coiled. The coil, perhaps 1.5 m in diameter and weighing 1000 kg, is laid flat on a horizontal rotary pallet table and the end fed into a reciprocating press with a set of progressive die cutters. These cut the strip and expand it in one operation, forming a double-width grid-form with a solid central section. As the expanded strip emerges from the press it has a typical curled or 'gull-wing' form, resulting from the cutting and pushing action of the cutting teeth. This is then fed into a shaper which flattens the mesh, partly rolling over the upstanding nodes in the process, ensuring that the grid-form is to the reqired thickness, and punching out areas from the central solid section, leaving opposing lug forms attaching the two halves of the strip. See Figure 3.7.

LEAD   EXPANDED   PASTED   PAPER
COIL   STRIP   SECTION ⌐   OVERLAY

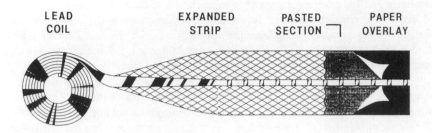

Fig.3.7. Expanded metal strip
(Based on a drawing from Cominco Inc.)

The continuous expanded strip is then fed into a pasting machine and, immediately after pasting, a thin paper material is rolled into both sides of the expanded ribbon. This is then cut into the final plate form before being fed through the flash-drying oven, and off-loaded to pallets for conditioning. See Figure 3.8.

In the second process the starting material is cast lead/calcium alloy strip, made by rotating a water-cooled drum dipping into a trough of molten alloy and peeling the resulting frozen strip off the drum (see Figure 3.9) and coiling. The molten metal, from a holding furnace, is continuously pumped through the casting trough, keeping the metal at a constant temperature and height relative to the drum.

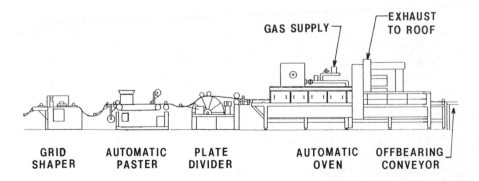

GAS SUPPLY ⌐   ⌐EXHAUST
                        TO ROOF

GRID   AUTOMATIC   PLATE   AUTOMATIC   OFFBEARING
SHAPER   PASTER   DIVIDER   OVEN   CONVEYOR

Fig.3.8. Schematic diagram of an expanded metal line
(Based on a drawing from MAC Engineering and Equipment Co. Inc.)

Fig.3.9. Continuous casting of strip
(Cominco Inc.)

An inert gas atmosphere is maintained over the molten metal both in the furnace and in the trough to minimise drossing. The strip is cast to the required thickness, needing no subsequent rolling or callendering operation prior to slitting. It is expanded by feeding under a rotating multihead slitter, which also forms the lug profile. The slit material is gripped at both edges by moving chains which expand the strip to a predetermined width (Figure 3.10.), before passing it through a set of adjustable rollers to ensure that the mesh is to a precise thickness. The strip is then pasted and cut in a similar manner to the previous process.

One advantage that this second method offers is that wire width and diamond size can be varied down the grid, and feet can be formed on the base of the grid form.

As noted above, both methods are in widespread use and can run at rates of up to 200ft/min. through the expanding, pasting and cutting operations. The ability to paste continuous strip rather than individual grids or grid multiples is a particular benefit. Production of the initial strip is somewhat slower, however, hence the need for a coiling operation. Most of the alloys in question require some age-hardening prior to forming the grid.

74

There are differences in the microstructure of the grid metal resulting from these different processes. In the case of the wrought strip a layered or laminar structure is formed as a result of the rolling process, whilst in the case of the cast strip the structure can change through the section of the strip because of differential cooling of the alloy on either side. Differences in corrosion behaviour have been noted due to these structural differences.

Fig.3.10. Rotary slitter-expander
(Cominco Inc.)

In principle there is no reason why wrought strip should not be expanded by slitting and pulling, or that cast strip should not be expanded by a reciprocating press.

However, particular attention needs to be paid to the characteristics of the alloys chosen, mainly lead/calcium of course, because of the performance requirements of the modern battery, although some low-antimony alloys, and even dispersion-strengthened pure lead, have been considered. The alloys used in each process are the result of a great deal of development to obtain the necessary combination of ductility, tensile-strength, narrow freezing-range, or other characteristics suitable for that particular process. It is therefore unlikely that one alloy will be equally suitable for both processes.

3.2.3 Continuously Cast Grids
A newer development is the direct casting of the final grid form, as strip, from molten alloy over a water-cooled drum into which the grid form has been cut. Immediately after casting, the grid strip is cooled in water and processed through tension rollers before coiling. The technology is very new and has yet to see widespread use. It has the advantage that virtually any design of grid can be considered.

3.3 Small Parts Casting
Although modern production methods have dramatically reduced the need for pre-cast interconnectors, most manufacturers have retained a facility to cast a variety of small pre-cast components, such as terminal inserts, interconnectors for small volume runs, lead bars for terminal fabrication and for connectors in plate formation, etc.

Very small volume requirements are still sometimes met by hand gravity-casting, but, in general, rotary multi-cavity gravity-casting machines, automatically producing a variety of components as demand dictates, are used.

Pressure diecasting is also used, particularly for items such as terminal inserts.

3.4 Paste Mixing
This is essentially a very simple process, consisting of mixing the leady lead oxide with sulphuric acid and water, to a consistency suitable for the pasting machine to apply it to the grid support. However, the formulation and mixing of the materials is quite complex. The physical behaviour of the paste, when subject to various stresses, and the chemistry of the paste-drying stage to follow, is as complex as that found in concrete technology.

Paste-mixing is carried out in a large flat-bottomed mild steel bowl, typically about 3 m in diameter, and capable of handling up to about 1500 kg of paste. (Figure 3.11). Mixing can be by conventional dough-type paddle mechanisms or by a pair of broad rotating wheels with horizontally inclined blades. Scraper blades continuously return material from the edge of the vessel. A quantity of water is first put into the mixer, followed by the correct amounts of lead oxide and sulphuric acid, (added slowly to ensure homogeneous mixing and to control temperature rise). Make-up water is finally added to obtain the correct consistency. Often chopped Terylene fibre is also added, which helps to bond the material together during its working life.

If the paste is to be used to make negative plates other additives are incorporated at this stage. These include barium sulphate, carbon black and derivatives of lignin sulphonic acid (which reduce the consolidation and blocking effects caused by cyclic recrystallisation). An exothermic reaction ensues, due to chemical changes in the lead-oxide/sulphuric acid mix, which requires the water jacket of the mixer, and air-cooling, to moderate it.

Fig.3.11. The paste mixer
(Oxmaster Inc.)

When paste mixing is complete, normally after about 20 minutes, the paste has a consistency like a moderately stiff cement mix, and should be used as quickly as possible, preferably within about 45 minutes; otherwise the paste begins to harden and set, and must be scrapped.

Working parameters used to check that the paste is of the right formulation and consistency are 'penetration', as measured by the depth that a pointed weight penetrates a sample of the paste, and 'cube-weight', the weight of a standard volume of the paste.

The paste at this stage consists of lead oxide, monobasic and tribasic lead sulphates, with a metallic lead content still at about 20%. Moisture content is of the order of 15%.

It is most important that the paste is as homogeneous as possible, and that mixing temperatures are controlled. Mixers should be thoroughly cleaned down between mixes, to ensure that lumps of old paste do not contaminate subsequent batches, and that additives used for negative mixes do not contaminate positive mixes (particularly barium sulphate).

It is also important that the water and sulphuric acid used in the factory for processing are free of harmful contaminants. Iron, manganese and chlorides can all lead to unacceptable levels of self-discharge in the finished battery. Water that is of questionable purity should be passed through a de-ioniser before use. Acid should be supplied to a specified purity, and should be stored in lead-lined or plastic tanks, and definitely not in unlined steel containers.

3.5 Pasting

The mixed paste is fed to a small hopper situated directly over the pasting head of a pasting machine, shown in Figure 3.12. This is conveniently done by arranging the paste mixers to be on a gallery over the pasting machines, so that the paste can be dropped under gravity in controlled amounts, but in many older installations paste is still moved by barrows and loaded by shovel. It is important not to overload the hopper as 'bridging' can occur, leading to incomplete filling of the grid. Older pasting machines apply paste to the grid from both sides while the grid is being transported downwards through a vertical orifice. However, it is difficult to process the very light grids in use today, and the flatbed paster in which the grid is carried horizontally through the paster is the one in most widespread use.

Cast double grids are stacked vertically on rails, supported by the lugs at either end of the double grid, and from here are picked individually onto a moving flat heavy-duty canvas belt, so that they are virtually abutting, and passed under the pasting head. The pasting head contains a series of coarsely-knurled rollers, contra-rotating, which pull paste in from the hopper and force it down, under considerable pressure, onto the grid, causing the paste to wrap round the side of the grid in contact with the belt.

Scraper blades ensure that excess paste is removed. This is returned to the hopper. It is important to ensure that plate lugs are kept as clean as possible for subsequent processing, and that plate edges are free of paste residues. The pasted grid, or plate as it is now described, is then transferred to a chain conveyor for flash drying.

Continuous grid material, either expanded, or as-cast, is driven through in essentially the same manner as above, and is cut to the final grid form immediately prior to the flash-drying oven.

3.6 Flash Drying

The pasted plates, supported horizontally on chains, are transported rapidly through an oven which exposes them to either gas or electric radiants, so that the surface skin of the plates is rapidly dried without removing moisture from the inside of the plate. This allows the plates to be stacked together for the subsequent curing or conditioning treatment, without sticking together, but does not remove the bulk moisture, which is vital for the correct chemical reactions to take place.

Fig.3.12. A conventional pasting machine
(MAC Engineering and Equipment Co. Inc.)

Double plates produced from the double cast grid described earlier are arranged in one of two ways as handled off the end of the flash-drying oven .They can be racked vertically on stillages, with each double plate supported by its lugs and a small space left between each double plate, or they can be stacked horizontally on slatted wooden pallets in bundles up to about 20cm high. The single plates resulting from the continuous strip methods can only be handled by stacking, as they do not have the opposite supporting lug. Advocates of the racking method maintain that, for the curing process that follows, the plates are in a much more even and controllable environment, while those that stack point to better use of space, and ease of handling. Both methods give satisfactory results provided that adequate controls are in place.

Flash drying reduces the moisture level to about 9% and at the same time the free lead drops to about 14%.

3.7 Curing

This is sometimes referred to as conditioning or hydrosetting.

Put simply, this is the plate-drying stage. It is also the stage where the final structure of the plate, which will define its performance in service, is developed by the rate and kind of chemical reactions that take place.

Stacked, or racked, plates are put into large enclosures, where the temperature and humidity of the environment can be controlled. A programmed sequence of initially high temperature and humidity, gradually reducing, for a total period of 72 hours (fairly standard) then ensues. Oxidation of the residual metallic lead particles in the paste takes place, giving a slow temperature rise, and a gradual drying-out of the plates. The rise in temperature, and gradual drying-out, cause a recrystallisation of the paste material to take place, with the formation of more complex basic lead sulphates. The proper progress of this curing is very important, in order to give a firm and strong working material in the plate and good adhesion to the grid material, thus ensuring a long battery life.

At the end of the 72 hours, plate moisture is down to less than 5%, with a free lead content below 4%. The plate feels dry and stiff, and there is a good grid/paste bond.

However, the material is still not in the final active form. The material in the positive plate must be converted into lead dioxide, and that in the negative plate to metallic lead.

The manner in which this is achieved, and the stage of manufacturing at which it is carried out, vary according to the subsequent storage requirements, and with each manufacturing method.

Essentially, the electrochemical processing required can either be carried out at this stage, resulting in dry-charged plates, ready for assembly into batteries, which are subsequently sealed, and distributed dry; or the existing plates are assembled into batteries, which then go through electrochemical processing, and are left filled with acid for subsequent distribution.

3.8 Plate Formation and Dry-Charge

3.8.1 Plate Formation
Conditioned double plates are placed in four-sided plastic crates, which have vertical slots in their long sides. Alternate positive and negative plates are placed in these slots (sometimes two or three plates per slot), in such a way that the lugs of positive plates are to one side of the crate, and the lugs of the negative plates are to the other side. Positive lugs are electrically connected, as are the negative lugs, forming in effect what will become a many-plated 2 volt cell. This is then placed in a vat of acid of a suitable concentration, interconnected with other like crates in series, and electrochemically charged, so that the positive material is converted to lead dioxide, and the negative material to porous lead.

Because the amount of material per plate has been specified for the plate ratios in the final battery, the positive to negative ratio in the many-plated formation vat has to be adjusted accordingly, to ensure that an imbalance in charge does not result.

When the plates are fully formed the crates are disconnected, lifted out of the vats, and the plates thoroughly washed in water to remove the sulphuric acid.

3.8.2 Dry Charge
The formed positive plate, consisting mainly of lead dioxide, can be dried in air. After washing, plates are hung, spaced out slightly, on racks, which are then put into a circulating-air oven at about 60°C for about 12 hours.

The negative plate, on the other hand, consists mainly of 'spongy' lead, a highly active form of the metal that will oxidise very rapidly in air when moist.

A variety of methods are in use for drying negative plates using either inert gas ovens, or vacuum ovens, or by heating rapidly between hot-plates.

Hot-plate drying, in which plates are placed between two metal plates at about 180°C and clamped for approximately 3-4 mins until dry, depends on the steam generated by the process to expel air until the plates are fully dry.

Inert gas, (normally hot, burned, heating gases) and vacuum ovens obviously have no air involved, and dry over longer periods, often similar to the positive plate drying time.

It is vitally important that the dried plates, either as plates or as dry-charged batteries, are stored in a dry atmosphere to prevent degradation. It is also important that the other components used in assembly are dry.

3.9 Parting
Before going into final assembly, plates made with cast grids, invariably cast as multiple grids (doubles for sizes associated with four-wheeled vehicles, but greater multiples for the smaller motorcycle sizes) must be cut, or parted.

This is a very simple mechanical operation that hardly warrants a heading of its own, but accuracy of cut, without distortion or other damage, is very important for subsequent operations. Poor handling can give rise to excessive scrap, and/or environmental problems in the form of lead dust.

Traditionally, plates were parted by hand along thinner weaker sections deliberately cast into the grid form. However, for economic and operator exposure reasons, this practice has been largely superseded by machines.

These normally rely either on rotating circular knife blades, or on sawing by either band or circular saws. Machines are boxed-in to contain the lead dust emanating from the cutting.

3.10 Assembly

Assembly methods vary enormously, from the highly-automated facility utilising the latest materials and techniques, through the hand assembly methods, albeit very advanced, used for the multiplicity of small motor-cycle batteries, to the low-volume hard-rubber containered batteries still required for some applications. For brevity, we will confine the description to the newer, high-volume techniques, in use for the majority of 12 volt car batteries (see Figure 3.13).

The initial step is to assemble the cell group, consisting of alternate positive and negative plates, interleaved by separators. The separator has, for many years, consisted of a semi-rigid plaque of a suitable insulating material, cut to just overlap the plate perimeter on all sides, in order to prevent edge shorting. However, recent developments have produced tough, flexible materials, capable of being wrapped around a plate, and sealed to provide an envelope.

Positive and negative plates, together with separators, are fed to the magazines of an automatic, pre-programmed, group stacking machine which picks plates and separators alternately, to the chosen configuration or, where an enveloper-stacker is used, one type of plate, previously chosen, is wrapped and edge-sealed in separator material, normally cut from a roll, as the picking operation proceeds. Figure 3.14 shows a typical machine.

Cell groups are conveyed to a cast-on-strap machine for connecting individual positive plates together, and negative plates together. Cell groups are inverted, the lugs brushed to remove traces of paste, fluxed, and lowered into moulds containing molten lead alloy. This is allowed to freeze off, and the cell group, complete with the connector strap, lifted, returned to the upright position, and dropped into a compartment in a battery container, the partition walls of which have previously been punched with holes in the correct position to facilitate intercell connection. Six such cell groups are required for a 12 volt battery, and it is normal to process all six simultaneously in this fashion. Figure 3.15 illustrates the stage of the casting on process where the packs have just been removed from the casting moulds in the cast-on-strap machine, while Figure 3.16 shows the finished pack ready for placing in the container.

Intercell connection, alternating positives of one cell to negatives of the next and vice versa, to produce a series connection, is achieved by electrical resistance-welding of adjacent cast-on-straps through the holes pierced in the partition walls of the container.

The upright parts of adjacent straps are clamped over the hole, the central parts squeezed into contact, and a current of some 7000-10000 amps passed, for typically about 5 cycles of 50 cps alternating current.

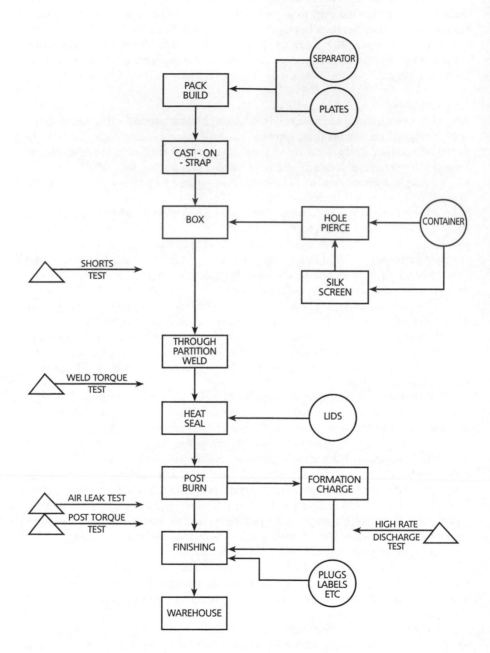

Fig.3.13. Assembly flow diagram

The end cell straps not connected to the adjacent cells have had pencil- like upright extensions cast at the cast-on-strap stage, in order to make the external connections through the lid, as shown in Figure 3.17.

The battery is now ready for lid sealing. This is achieved by melting peripheral, and partition, abutting surfaces of both container and lid, by contact with a hot-plate, held at approximately 125°C, for five seconds, and then quickly clamping the lid and container together until cooled, typically another five seconds, having ensured that the upright strap extensions, described above, are in correct position, through the lead terminal inserts previously moulded into the lid.

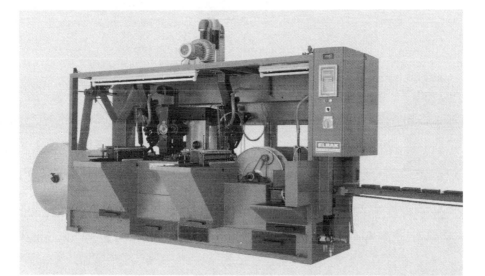

Fig. 3.14. A typical enveloper/stacker
(Elbak Batteriewerke GMBH)

Finally, these extensions are fused, or 'burned', to the inserts, to form the terminal posts. This is normally done by using a very small gas flame, either manually or automatically, but other welding methods are to be found. The 'lead-burning' operation, used at various points in the battery manufacturing process, differs from conventional welding in that no flux is used, no inert gas cover is required, and the welding rod used is the same as the base material.

At this stage all the joins have been made between container and lid, and it is important to ensure that there are no leaks, either from cell to cell, or round the periphery of the lid. This is normally achieved by having an automatic air-leak tester at the end of the line, which pressurises alternate cells, and senses any pressure drop. Typically a pressure of about 3 psi is used. Batteries that pass can be automatically hot-branded with a type code, date brand, and inspection stamp.

Batteries made with dry-charged plates are to all intents and purposes complete, with vent plugs, sealing tape, final labelling, and perhaps cartonning, being applied at the end of the line.

Batteries made with dry unformed plates are ready for formation-charging. In this state they can be stored without deterioration for long periods.

Fig.3.15. Packs with connectors in the cast-on-strap machine
(T.B.S. Engineering Co. Ltd.)

Fig.3.16. The cell pack with interconnectors cast-on
(Lucas Yuasa Batteries Ltd.)

3.11 Formation Charge

It will be recalled that the chemical structure of both positive and negative plates, as processed prior to this stage, has to be changed in order for the battery to perform its function. The complex oxides and sulphates have to be transformed to spongy lead at the negative plate, and to lead dioxide at the positive plate.

This is achieved by a straightforward, but controlled, electrochemical charging process, the detail of which varies greatly from manufacturer to manufacturer. Essentially, the dry battery is filled with sulphuric acid of a predetermined concentration, allowed to stand for a short period, and then charged through a pre-programmed series of constant-current levels, until full charge is achieved.

Most manufacturers opt to fill with an acid concentration of 1.23 specific gravity, in order to achieve a final specific gravity of 1.28, at which most batteries go into service in temperate climates (sulphates within the plates add to acid concentration during charge, which, depending on the paste formulation used, will normally amount to about 50 points in gravity), without need for adjustment, which can be a messy process. However, such a concentration of fill-acid increases the initial temperature of the battery, due to the exothermic reactions which take place between the acid and the plate materials. Too high a temperature at this stage, or later in the charge, can lead to subsequent plate-shedding problems, and an upper limit of about 65°C is normally considered safe. This is achieved by spacing batteries on the charging platform, to allow for good air circulation, or placing them in water baths. Batteries are also allowed to stand for a period of perhaps one hour, to allow the initial temperature to subside. Allowing too long a stand-time at this stage can lead to other undesirable 'sulphation' reactions taking place.

Fig.3.17. The battery prior to lidding
(Lucas Yuasa Batteries Ltd.)

The other problem encountered with first filling with a higher concentration of acid is that charging efficiency is reduced, leading to wasted energy, and a longer processing time. However, using acid of a lower gravity means that the final gravity is low, and a tip and refill, followed by further charge to finish the process, are required. Hence the preference for the 'one-shot' process.

Typically, the process is designed conveniently around a 24hr cycle, with a 40Ah (20hr rate) battery having a regime of a one hour initial rest, four hours charging at 4A ($0.1C_{20}$), twelve hours at 8A ($0.2C_{20}$), and four hours at 4A again. Given good air circulation, such a regime would keep the temperature well within control, and also allow full charge to be reached within a 24hr turnround cycle.

Specific gravity and temperature are monitored throughout the process. An idea of the inefficiency of the process can be gained from the above figures. The total input ampere-hours represent about 3.5 times the capacity gained.

It is usual to batch batteries through the process using a series/parallel arrangement. Direct-current power supplies, operating at voltages of up to about 100V are generally used, with voltages above this level being inadvisable where the safety of operators is involved, particularly bearing in mind the wet, acidic, environment often found. Hence, with charging voltages reaching in excess of 15V, a maximum of 7 batteries can be charged in series (see Figure 3.18). Depending on the constant current output of the power supply, a number of parallel legs of series connected batteries can be linked. Perhaps the most difficult part of the process is to ensure the continuity, and balance, of these parallel arrangements throughout the charging time. Electrolyte specific gravity and temperature are monitored at regular intervals. Automatically monitored facilities can be allowed to operate at much higher voltages.

At the end of the charge, batteries have their acid levels adjusted, before being washed, dried and finished ready for despatch.

3.12 Separators

A good separator must meet the following criteria :

Low mean pore size, to allow free passage of ions, but prevent 'treeing'.

Low pore size range, to evenly distribute porosity, and therefore current flow, over the whole area.

High total volume-porosity, to maximise total ionic conductivity. (In addition, it must be remembered that the third element involved in the electrochemical reactions is sulphuric acid, and the more volume that is occupied by the separator material the more acid is displaced, and unavailable for reaction.)

The separator must obviously be chemically stable in the presence of sulphuric acid, and must be resistant to the highly oxidising effects of lead dioxide on the positive plate. It must be chemically and mechanically stable over the whole temperature range in which it operates.

The modern separator must be chemically 'clean', i.e., it should have no impurities which might be leached out and subsequently impair the function of the battery.

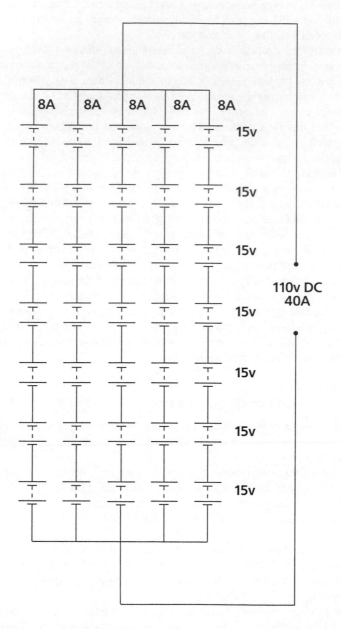

Fig. 3.18. Series/parallel formation-charge

It must have the toughness (with rigidity in the case of plaque separators, or with flexibility in the case of envelope separators) to cope with the rigours of production handling, and the compressive and vibration stresses in service, which can be very high due to vehicle vibration.

Over the years many materials have been used as separators, including wood veneers, rubber, PVC from emulsions, PVC as a sinter, cellulose (paper), polyethylene, glass mat, and phenolic resins.

In recent years polyethylene has started to dominate the scene, with cellulosics still in use as a lower cost (and lower performance) alternative, and glass mats used for motorcycle applications, where the compressibility and vibration resistance are vital factors, and the more specialised mats used for the sealed lead-acid batteries.

Polyethylene separators are formed by making an emulsion of high density polyethylene with water and a pore-forming material, such as starch or oil. When the emulsion is spread onto a stainless steel band, and passed through a heated oven, the water evaporates, leaving a thin polymeric film, with a certain proportion of pores, already ribbed after passing through profile rollers. When the film is subjected to a hot acid bath the starch is hydrolysed out, leaving a microporous strip, which, after washing and drying, is ready for use.

3.13 Containers and Lids

The ubiquitous and traditional black box of the lead-acid battery, usually made of hard rubber or resin rubber, has largely disappeared in recent years, giving way to modern plastic containers and lids, often brightly coloured.

The most favoured plastic is polypropylene, usually copolymerised with some polyethylene. The development of such materials for containers brought about perhaps the single biggest performance and life improvements to the automotive lead-acid battery. They are far superior to the older container materials in terms of toughness, impact-resistance, thermal shock-resistance, density, ease of production (after a lot of initial development), chemical inertness and purity, and, of course, are thermoplastic, i.e., can be melted and resolidified without degradation, and are therefore heat-sealable.

These properties enabled designers to significantly reduce wall sections, and therefore weight and size. The ability to make sealed intercell connections through the partition wall reduced the lead mass required for such connections, and at the same time reduced the electrical resistance of the interconnections. Heat sealing of the lid to the container eliminated the need for a sealing material, and improved the reliability of the seal. For a conventional car battery of 40Ah this represents a saving of about 15% by weight, and 9% by volume, although, of course, as size increases these percentages reduce.

The inclusion of some polyethylene in the plastic formulation imparts improved cross-linking, which increases the impact-resistance of the material.

Components are made by pressure injection-moulding machines, which, over the years, have become highly automated and reliable. Cycle times are of the order of 40-45 secs., often from multi-cavity moulds. Pre-cast lead terminal inserts are automatically fed into position in the mould used for lid production.

Natural polypropylene is translucent, which is advantageous in that the acid level can be seen through the container wall. The material can also be pigmented to provide an enormous variety of coloured components, providing the marketing department with opportunities for differentiating products.

Vent plugs, manifolds, handles and terminal covers can also come from different grades of the same plastic.

3.14 Sulphuric Acid

Sulphuric acid is a clear, colourless, odourless liquid, which in its most concentrated form, about 98% pure, is quite viscous, almost non-conducting, and highly corrosive as an oxidising chemical, and dangerous to handle. It has a specific gravity of 1.84 (15°C). It is very hydrophilic, and combines with water, generating a great deal of heat. Indeed the reaction is so violent that the water should never be added to the acid, but the acid added slowly to water.

Sulphuric acid is made by the 'Contact Process', in which sulphur dioxide is oxidised to sulphur trioxide, over a catalyst.

Sulphur, either in elemental form, or as a metal sulphide, is heated in air to produce sulphur dioxide. This is purified, mixed with an excess of oxygen, (stoichiometrically greater than that required to oxidise to sulphur trioxide), and dried, normally over sulphuric acid, before passing over a catalyst at a temperature of about 400°C. The resulting sulphur trioxide is then absorbed in a stream of fairly concentrated sulphuric acid, cooled to about 70°C.

For many years vanadium pentoxide was the standard catalyst used, but in recent times more advanced materials, based on other vanadium and potassium salts, have taken over.

Sulphuric acid is a widely available, and inexpensive, reagent used in a variety of chemical processes. The specification of battery-grade acid must be tightly controlled to avoid harmful impurities. This applies to both acid and mixing water, and for ease of storage, is normally supplied at a specific gravity of about 1.4-1.6, with a final acid gravity correction for paste mixing, or charging, being carried out in the factory. It is normally stored in lead-lined tanks, or reinforced plastic tanks. Transport cost, particularly for dilute sulphuric acids, is a significant proportion of total cost, and it is normal to source as locally as possible.

3.15 Lead

In comparison with the most abundant metals, aluminium and iron, lead is a rare metal. Even copper and zinc are more abundant by factors of 5 and 8. However, lead is found in many parts of the world, with the USA and Australia being major producers, and known deposits provide reserves for many hundreds of years at current levels of usage. Indeed, the significant decrease in the use of tetraethyl-lead as an additive for petrol, and the decrease in the use of lead compounds in the paint industry, coupled with spreading legislation requiring the recycling of batteries containing lead, has brought about a major decrease in primary lead-mining demand.

The most important ore is Galena, lead sulphide, often found associated with the sulphides of zinc and silver. In fact, the economics of extraction, and refining, of lead is very much affected by the demand for these other metals, and the market prices are closely interrelated.

3.15.1 Smelting
The mined ore is crushed and separated from the rock and graded into its main constituent minerals by a flotation process.

The separated ore is sintered at about 600°C, together with smelter by-products, and some coke, which converts the sulphides to oxides, driving off sulphur dioxide and other gases.

The sinter, with coke, silicates and lime, is then smelted in a blast furnace operating at a temperature which can reach 1200°C in its hottest part, and molten impure lead is tapped off at the end of this process.

3.15.2 Refining
The refining process consists of a number of sequential chemical stages to remove the impurity metals.

In a reverberatory furnace, held at about 700°C, antimony, arsenic and tin are oxidised and removed.

Zinc is added to the melt, which forms intermetallic compounds with the noble metals present. Silver is the main metal removed by this reaction, and hence the process is known as 'desilvering'.

The excess zinc is then removed by a variety of techniques: oxidation by chlorine gas, vacuum extraction, or oxidation using caustic soda.

Bismuth is then removed by adding lead-calcium alloy, and magnesium metal, to form a complex intermetallic compound with the bismuth, and finally the excess calcium and magnesium are removed by chlorination, or by caustic soda.

3.15.3 Recycling
As mentioned before, a significant proportion of the lead feedstock for the battery industry is secondary, or recycled, lead.

The economics of recycling are critically dependent on the world price of lead, as this influences the value of scrap, which in turn must more than cover the costs of collection and transport. (The market price for lead is made on the London Metal Exchange, or LME, where all the major contracts are transacted. The price is set by the market several times each day.) Legislation, however, is forcing the battery industry towards recycling, irrespective of economics, so that the cost of recycling will inevitably form part of the product cost in the future.

The spent battery represents a relatively concentrated and pure source of lead, with assay values that will not change by more than a few percentage points. Hence the reclamation process can be relatively straightforward, although the mix of lead-antimony and lead-calcium alloys present in a modern pile of spent batteries has presented the recycler with some additional problems (see Section 3.2.1).

In very large installations blast-furnace technology is still used, with whole batteries as the feed. Reverberatory furnaces are also in use, but require a finer feedstock. Perhaps the most widely used, and convenient, technique is the rotary furnace, due to its flexibility.

Modern facilities include a first-stage automated break-up step, from which the polypropylene case material can be extracted, and then reclaimed. Cell groups, consisting of grid metal, lead, and lead sulphate paste, top lead parts, and separators, then form the main feedstock for the furnace, together with controlled amounts of coke and lime. The refining stages are somewhat different to those for primary lead as, of course, the smelted metal contains few of the natural ore impurities, but relatively high quantities of added alloying metals, including antimony and calcium, the former of which is needed less and less by the industry.

Battery recycling is dealt with in greater detail in Chapter 4.

3.16 Lead in the Factory

Lead in its various forms is known to be poisonous. It can accumulate in the body, and is slow to be eliminated. It can reduce foetal growth, and cause premature birth. It may have an effect on brain function in small children, and in adults it can affect the nervous system, and renal function.

Hence, it is important that measures are taken to protect people working with lead, and to contain the spread of the material.

Within the battery factory, problem areas are those associated with lead fume arising from hot molten-metal processes such as casting and joining, and dusts from lead oxide production, paste mixing, pasting, and the handling of dry plates.

Lead fume is relatively easy to contain, extract, and condense. Problems are only likely to occur when the hazard is not recognised. The dusts of lead oxide and other lead compounds pose some problems, however. Although the materials are heavy, particles can be very small and hence quite mobile, whereas large particles can be difficult to extract. Containment in the oxide manufacturing process is normally good, in that a largely closed automated system is normally found. Paste-mixing and pasting processes invariably rely on frequent wetting-down to overcome the problem. Handling of dry plates through parting, transport between processes, and assembly, are the most troublesome, requiring good cleaning methods, good work-station design, good extraction, and generally good housekeeping.

Personnel exposed to the hazard should wear protective clothing, including a face mask if necessary; and rules prohibiting smoking and eating within the working environment normally apply. Protective clothing should be removed on leaving the working environment, and personal hygiene should be of a high standard. Employee education is vitally important.

Dust sampling at strategic points through the factory is used as a control, as is personal sampling of the breathing zone of high-risk operators, and regular inspection and testing of extraction systems. It is usual for all personnel working in lead areas to have regular medical checks, including blood testing.

Filtration systems have to be of a high standard, to limit the escape of airborne exhaust dusts, and liquid effluent plants must be capable of trapping lead sediments, as well as neutralising acid wastes.

3.17 Economic Considerations

Until fairly recently battery manufacturing methods varied significantly from one maker to another. Among the larger manufacturers, technology was largely home-grown, leading to the fabrication of machines, tooling and equipment 'in-house'. The larger manufacturers also had a high level of vertical integration, often with smelters, container moulding and separator manufacture.

As the market has grown, competition has become more intense, and development costs have soared, so that this situation has changed dramatically,as explained below.

3.17.1 Plant and Equipment Supply

Key items of manufacturing plant are now developed and produced by a relatively small number of specialist machine-tool engineering companies almost solely dedicated to the lead-acid battery industry. These are mainly concentrated in the USA and Europe, with Japan and the Far East yet to have an impact.

This has tended to lead towards a greater uniformity of production methods.

3.17.2 Raw Materials and Component Supply

Lead, of course, is the main raw material, and as noted before, its price is set globally on the LME(London Metal Exchange). Hence, although contract arrangements between supplier and customer will vary, particularly regarding process and transport costs, the base-price movements will be common. These commodity price movements, influenced often by factors outside the normal supply and demand situation, can have a serious effect on the economics of the battery business.

The manufacture of separators is now largely in the hands of a relatively few specialist separator manufacturers, where the economics of scale outweigh the possible advantages of control that individual battery manufacturers may have had. The separator producers now have their own R & D facilities, and compete with each other for technical and competitive advantage. The product is available to all users.

Polypropylene containers and lids, for similar reasons, tend now to be made by specialist moulders, albeit often with tooling owned by individual customers. Even when this is the case the tool design and fabrication expertise is normally in the hands of the moulder.

3.17.3 Battery Manufacture

As in many other industries the development of electronics, sensors, and actuators has provided the means for more and more automation. Productivity gains in recent years have been immense. The industry is no longer labour-intensive as it was for many decades, but has become capital-intensive.

The modern battery plant is a high-volume, high-productivity unit, where the emphasis is on maximising output, minimising downtime through breakdown or changeover, and achieving a high level of consistency in the product.

The much smaller number of operators required must be much more technically skilled and adept than their forebears.

Modern vehicle developments have placed ever greater demands on the battery, which economically can now only be met by production methods using the latest materials and technology.

The economies of scale have had a major influence in this situation and, in recent years, through merger and acquisition, manufacture has become concentrated into a quite small number of very large, often international, groupings, able to maintain extensive R & D facilities, and having the market to afford the capital investment needed to keep abreast of the competitive situation. This aspect is described in greater detail in Chapter 4.

3.17.4 Distribution

The average car starter-battery is still a bulky, heavy, and difficult item to transport, particularly in wet-charged form. It contains highly corrosive sulphuric acid, which in the event of accidental damage, can leak out. It does, when freshly charged, produce some gases, including hydrogen, which in certain concentrations is explosive. Indeed, gases can continue to be evolved at low levels, through self-discharge reactions, for long periods. If shorted, it can produce about 3kW of power, albeit only for a short period. By this measure a container-load of batteries, weighing some 30tonnes, has the potential for producing 10MW of power!

For these reasons, in most of the developed countries, legislation exists regulating the transport of batteries in bulk by road, rail, air and sea. Regulations restrict, and invariably lead to an increase in what is already a proportionately high transport cost, due to the unit weight and volume. (Similar regulations also apply to the spent battery requiring recycling.)

This tends to limit the distance over which wet-charged batteries are distributed. The dry-charged battery is more expensive to manufacture, but when distributed over longer distances can become economic.

The wet-charged battery has a limited shelf-life and therefore tends to be used for fast-moving product, whereas the dry-charged battery has a much-extended shelf-life, but both expertise and additonal expenditure of time by the retailer are required to activate the battery.

CHAPTER 4

The Battery Industry and its Environmental Impact

4.1 Some Considerations of the Structure of the World Battery Industry

The world battery industry is very large, and is particularly complex in that it varies very much from one region of the globe to another in its characteristics such as its structure, size, market shares and manufacturing methods.

There are three main types of battery in general use, namely dry batteries, nickel-cadmium and lead-acid batteries, although there are scores of other specialised types, such as sodium-sulphur, silver-zinc, nickel-zinc and so forth, made and used in very small quantities for very specific purposes.

It is difficult to compare and contrast the size and importance of the markets for these three different main classes, because they are so different in size, weight content and value, and global statistics are very hard to obtain, particularly for the dry battery market.

Nevertheless, in order to give some perspective view of the importance of the lead-acid automotive battery market within total world battery markets, approximate volume and value figures for each class of battery are given, and it should be stressed that these are indeed best approximations.

4.1.1 Dry Battery Market

Approximately 20,000 million units per year are made world-wide, with a total value of $6,000 to $7,000 million per annum. There are three main types:

> *primary or disposable* (alkaline or zinc), for torches, toys, and all manner of low-voltage electrical/electronic appliances,

> *secondary or rechargeable* cells, for similar applications to those above, but where customers will pay much more for the recharge capability, and

> *button* cells for watches, hearing aids, etc.

Of these about 50% are alkaline, 40% zinc, 6% button cells (mainly lithium types), and the rest rechargeable or miscellaneous.

4.1.2 Nickel-Cadmium Market

Nickel-cadmium batteries are secondary or rechargeable, whilst, as mentioned above, over 90% of dry batteries are primary or disposable. Whilst nickel-cadmium batteries can be much larger and heavier than the normal dry battery, and indeed, in some applications such as ships' standby systems, a single battery can weigh several tons, most nickel-cadmium cells fall into the weight range of several hundred grammes to several kilogrammes.

Whilst there is a much closer parity in size and volume between nickel-cadmium and lead-acid batteries, the former are roughly five times more expensive than the latter, compared on an ampere-hr capacity basis.

There are two main types of nickel-cadmium battery in use:

small sealed cells for domestic appliances, commercial or military use, office equipment, electronic devices, etc.,

large industrial cells for trains, ships, buses, standby systems, etc.

It is estimated that current world production of sealed cells is about 750 M cells p.a., and of the much larger industrial units, about 40M p.a. to a total value of about $1,500 M p.a.

4.1.3 Lead-Acid Battery Market

There are three main types of lead-acid batteries:

SLI (explained below),
Industrial, and
Consumer.

The largest market for lead-acid batteries is as SLI (starting, lighting and ignition) batteries for automobiles and commercial vehicles, either fitted as original equipment or supplied as replacement batteries, and including also batteries for tractors and motor-cycles together with a number of sporting and general utilities such as golf carts, power boats, grass-cutting machines, invalid carriages and so forth.

The automotive sector is much the most competitive, with the bulk of supplies going to the vehicle manufacturers or the price-conscious replacement market, so that it is the sector of high volume and low margins.

Industrial batteries are generally much bigger and heavier than automotive batteries, and indeed, like nickel-cadmium batteries, single batteries can weigh several tons. They are used for applications such as milk-floats, fork-lift trucks, power station standby, emergency systems for hospitals and offices, and many more.

Consumer batteries are chiefly small sealed or non-spill types for portable electronic/electrical equipment, leisure use and so forth.

Approximate production figures world-wide (6 cell batteries) are:

Automotive	270 M p.a.
Industrial	10 M p.a.
Consumer	25 M p.a.

(In order to compare unit figures for lead-acid with unit figures for nickel-cadmium, the above figures should be multiplied by about factor six, since lead-acid units are given as 6 cell batteries. Also note that whilst automotive units are given as 12 volt/ 6 cell batteries, industrial battery installations are usually built up to higher-voltage units from single cells, and consumer batteries are normally 6 volt/ 3 cell units. The above figures have been normalised for this effect.)

Tables 4.1 and 4.2 summarise much of these considerations.

Table 4.1. Western World battery production 1990
(Excludes former Eastern Bloc countries) ($M)

	Lead-acid	Nickel-cadmium	Dry batteries	Total
Automotive	6800	-	-	6800
Industrial	2200	350	-	2550
Consumer	500	1200	5950	7650
Total	9500	1550	5950	17000

Table 4.2. Lead-acid use by market 1990 ($m)

	Western Europe	USA	Japan	Other	Total
Automotive	1860	2880	840	1220	6880
Industrial	1220	740	160	80	2200
Consumer	120	180	150	50	500
Total	3200	3800	1150	1350	9500

(Source for Tables 4.1 and 4.2: M.G. Mayer L.D.A. Report, May 1990)

Output of SLI batteries has risen steadily with the continuous rise in the world vehicle population, with production in 1995 forecast to be about 267 million units. About 24% will be used as original equipment and 76% as replacement batteries. In addition, it has recently become a quite dynamic industry, after several decades of slow change, in the matter of mergers and acquisitions, particularly in Western Europe, and earlier, in North America, as well as experiencing a recent quickening of the pace of improved technology and better production methods, which are dealt with in other chapters.

Thus it is only possible to give a 'snapshot' approach here, in the knowledge that, in the near future, further rationalisations and changes in structure will be taking place, so that a number of details and figures will have changed again.

The battery industry is the largest user of the word's lead supply. From a recent annual production rate of just over 4 million tons, the battery industry consumed just over 2.5 million tons, or 63%, compared with all the other uses such as cable sheathing, roofing lead, shot/ammunition, and so forth.

Table 4.3 shows the growth, in millions of units produced p.a. of the world's automotive battery market, over two decades. Within this context, Table 4.4 shows the approximate annual production, by various areas of the world, in 1991.

Table 4.3. World automotive battery market
(millions of units) (excludes motorcycle batteries)

1975	125	1986	205
1980	160	1987	208
1982	172	1988	213
1983	179	1989	223
1984	193	1990	234
1985	201	1991	239
		1995 (F/C)	267

(Source: Battery Council International Annual Statistics)

Table 4.4. Annual production figures by region
(millions of units)

North America	81
Western Europe	50
Asia/Pacific	58
Latin America	19
Africa/Middle East	18
Russia/Eastern Europe	22

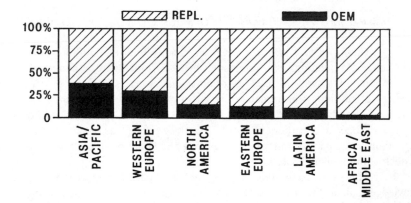

Fig.4.1. 1991 regional market split for OEM (original equipment manufacture) and replacement batteries

The characteristics of the markets tend to differ from one region to another. For example, Figure 4.1 shows the variation by region in the proportion of Original Equipment units to Replacement units.

This is partly accounted for by the degree of automotive manufacturing in the area, but also by the total vehicle population, battery life and type of battery usage. Figure 4.2 shows how much battery life can vary from one region to another.

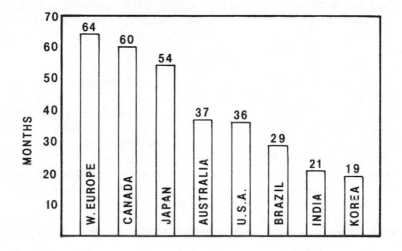

Fig.4.2. Estimated battery life by area for 1989
(Source: Battery Council International Statistics, The Battery Man, July 1993,p.21)

The significant differences in battery life shown here depend on a number of factors. The greatest is probably operating temperature, where the higher the under-bonnet temperature, the shorter the life. For example, battery life in the hotter Southern states of the USA is approximately half that of the figure for the cooler Northern states of the USA. Another important factor is the type of battery usage. In the less-developed countries, with a high proportion of rough roads, the batteries are subjected to a high degree of vibration, encountering forces as high as 10g, which can quickly cause deterioration of the internal components.

Also, the intrinsic quality and internal construction of the battery will have a very important effect on battery life, as explained in Chapters 2 and 3. These features will naturally vary considerably depending on the country of manufacture.

Consequently, the size of the replacement market in each area will be very dependent on these considerable differences in battery life.

The structure of the industry in each area also tends to differ, to some extent due to population density and geographical/logistics reasons. Lead-acid batteries are relatively heavy and bulky items, and not of high individual value, so that shipping them over long distances is not normally attractive. Also, when shipped in the 'wet-charged' state, they contain a considerable quantity of fairly concentrated sulphuric acid so that safety precautions must be in place in case of an accident. Batteries can be shipped in the 'dry-charged' state, in which they contain no sulphuric acid, and can be rapidly activated when required, but this entails more expensive production methods initially.

The European market has, over the years, tended to be very fragmented, and it has been recently estimated that there are over 200 battery factories in this area. In the UK alone there used to be over 50 factories, many of them small local units using hand-methods, but over the last decade or so these numbers have markedly reduced, with a strong trend towards mergers and acquisitions.

Several influences have caused this trend, notably, more stringent health and safety regulations controlling the handling of leady materials, an increasing rate of technological change, the increasing demands from vehicle manufacturers for improved product performance and quality, the increased purchasing power benefiting large groups, and the increased capital spending capability required from battery manufacturers who have to invest in the new improved plant and equipment.

For example, in Western Europe, whereas in about 1985 there were 8 manufacturers who controlled about 47% of the automotive battery market, by about 1993 only 2 firms controlled 48% of the market. The four largest firms were Varta-Bosch, CEAc-Marelli, Tudor (Spain) and BTR/Hawker, who together held about 72% of the market. However, at the time of writing, further rationalisation has taken place, with Exide of the USA having acquired control of Tudor Group, CEAc-Marelli, Dagenite and BIG Batteries in Europe, making them the largest lead-acid battery manufacturer in the world.

This trend in consolidation, which will continue steadily, has also been matched over the years by a parallel improvement in battery technology, manufacturing methods, improved materials and therefore battery quality. The last twenty years have seen a move away from a high proportion of hand-operations (once deemed necessary because during a large part of the production process the battery plates, as explained earlier, have a consistency akin to dry biscuits), to a much higher degree of automation and flow-line production, made possible by the much more sensitive equipment now available, and also, as explained earlier, by more stringent health and safety regulations.

It is generally accepted that the minimum economic size of a battery production unit is now about 2 million units per annum, and some sources claim it needs to be as high as 5 million units per annum.

Whereas twenty years or so ago many of the larger battery manufacturers built much of their own machinery and plant, today there are many specialist equipment suppliers who do not manufacture batteries themselves, but who can provide state-of-the-art plant or machinery, and these are described in Chapter 3. This trend has been partly caused by the increasingly sophisticated manufacturing equipment needing a high degree of specialist skills in its construction, partly by the higher degree of capital investment required in the machine-building facility, and partly by the fact that the increasing volume of battery output has allowed these specialist machine-makers to realise a sufficiently large market.

In North America, the market is much less fragmented than in Europe, with a greater concentration in large manufacturers, and also showing a recent tendency to acquisitions and mergers. In addition it has the unique feature that one battery manufacturer, Delco-Remy, with an annual production of approximately 12 million units per year, is owned by a vehicle manufacturer, namely General Motors. The large independent manufacturers are Exide (about 30M p.a. plus), GNB (about 20 M p.a.), and Johnson Controls (14M p.a.). Others include East Penn (5M) and Douglas (5M), with a few more at about 2M units per annum.

In the Asia/Pacific area, the Japanese market is very orderly, with a structure which has not changed much in recent years, and with production concentrated in five large companies, namely Japan Storage Company, Yuasa Battery, G S Batteries (Hitachi), Panasonic (Mitsubishi) and Furakawa. In other parts of S.E. Asia, such as Korea, the Philippines, Malaya and Australia, strong competition has applied recently. For example, when protectionist duties were phased out in the 1980s, Australian producers quickly lost more than 30% of their market to imports. One Australian producer, Dunlop Pacific, took strong measures to combat this situation, and, by acquiring GNB in the USA, has remained a 'World Class' manufacturer.

Somewhat in contrast, in Latin America, Africa/Middle East and Eastern Europe, production units tend to be much smaller, and production methods are manual rather than automated, which is understandable bearing in mind the lower production volumes involved. An exception to this rule is Brazil, which has the largest market in this group of regions, and where more highly automated production has been adopted.

In general, most battery manufacturers are in a number of different markets, including traction, stationary/standby, dry-cell, etc., as well as automotive batteries. Purchasing power for raw materials and components is maximised, some production processes apply to several types of battery, and profit margins on other types of battery are higher than those for automotive batteries, and hence help to support the latter.

Turning now to the relationship between the vehicle manufacturer and the battery manufacturer, the latter is usually responsible for the design, performance and quality of the item, as with most other vehicle components, except for the special case in N. America, where General Motors produce their own batteries.

The vehicle manufacturer will normally spell out the parameters required from the battery, such as engine-cranking performance, lighting-load capability, ability to supply power for various electrical equipment on the vehicle, and capacity to provide the low-level standing drain of security systems, etc., as well as the space available in the car to house the battery. From these and other requirements, the battery manufacturer will design and develop a unit to carry out the required functions, and these aspects are covered fully in Chapters 2 and 3.

In summary, the world-wide battery industry is a very complex and diverse activity, but it is dynamic and evolving quite rapidly. The last two decades have seen a marked improvement in battery performance, as regards battery life, power output per unit volume, size and weight reduction, and improvements in other features, at the same time as improved cost-effectiveness. These benefits have been brought about by the use of better materials for the containers, grids, separators and other components, as described in detail in Chapters 2 and 3, but also by much improved manufacturing processes, also described in these chapters.

4.2 Batteries and the Environment

4.2.1 Introduction

Since lead-acid batteries contain considerable amounts of lead and lead compounds, and since lead and many lead compounds, when in a finely-divided or dusty form, are physiologically harmful if inhaled or ingested above certain levels, it is obvious that care needs to be taken at certain stages of their life, when handling them.

It is fortunate that, during most of their life, when they are installed on a vehicle, or being used on standby service, or powering a milk-float, lead-acid batteries are quite innocuous from an environmental point of view, since their contents are retained within a strong, durable container. It is only during a part of the original manufacturing process, as described in Chapter 3, but even more particularly at the end of their service life, when they have to be disposed of, that they become a potential hazard. Such a hazard would ensue if discarded batteries were, for example, dumped on a land-fill site and the containers were broken. In this case, when the lead compounds dried out, they would become a hazard to anyone who disturbed the site and inhaled or ingested the resulting dust.

4.2.2 Physiological Hazard of Lead Compounds

Lead is a versatile metal which has been used for several thousand years in a wide range of important applications, but nevertheless it is also a toxic metal which, if absorbed into the body in considerable amounts, either through a single high exposure, or through long-term chronic exposure to lower amounts, can result in adverse health effects.

Once it has been absorbed into the body, lead is transported by the blood and either excreted or stored in various organs, predominently in bone. Generally speaking, low levels of exposure can be tolerated by the human body with no measurable impact other than an increase in the level of lead circulating in the blood-stream, or slight changes in specific biological compounds in the blood-stream. It is only at much higher levels of exposure (where lead-in-blood levels rise above about 100 micrograms/dl), that serious and clinically significant effects begin to develop, including anaemia, colic, muscle weakness and kidney impairment.

Such effects are virtually unknown in modern industry, or in the population at large, thanks to the health and safety measures which are in place as part of Environmental Controls, which are dealt with below. Ongoing measurements in the vicinity of lead-acid battery factories show that, thanks to the efficiency of modern containment and filtering methods, the lead-in-air concentrations prevailing in the surrounding atmosphere arises principally from leaded gasoline in cars rather than from the factory.

Table 4.5. Blood lead limits for occupational exposure (women)

Maximum lead level (μg/100ml)	Country
45	Germany Netherlands
40	Australia (parts) South Africa United Kingdom

(Principal Source; International Lead and Zinc Study Group, 'Environmental and Health Controls on Lead' (1989))

Table 4.6. Blood lead limits for occupational exposure (men)

Maximum lead level (μg/100ml)	Country
80	Australia (parts) Canada (parts) Ireland South Africa
70	Australia (parts) Belgium Canada (parts) EEC France Germany Italy Spain
60	Denmark Japan Morocco Netherlands Norway Peru
50	Canada (parts) Sweden USA
40	Finland

(Principal Source; International Lead and Zinc Study Group, 'Environmental and Health Controls on Lead' (1989))

Table 4.7. Air lead limits for occupational exposure

Maximum lead level (mg/m^3)	Country
0.20	Morocco
	Peru
0.15	Argentina
	Australia
	Belgium
	Canada
	EEC
	France
	India
	Ireland
	Italy
	Japan
	Mexico
	South Africa
	Spain
	United Kingdom
0.10	Denmark
	Finland
	Germany
	Netherlands
	Sweden
0.05	Norway
	USA

(Principal Source: International Lead Zinc Study Group, 'Environmental and Health Controls on Lead' (1989))

Table 4.8. Airborne lead emission limits

Maximum lead level (mg/m^3)	Country
50	South Africa
14/29	Canada
10-30	Japan
10-20	Spain
10	Australia
	Austria
	Germany
	India
	United Kingdom

(Source: International Lead and Zinc Study Group, 'Environmental and Health Controls on Lead' (1989))

Table 4.9. Typical effluent limits for lead

Maximum lead level (mg/l)	Type of discharge	Country
0.05	inland waters	Australia
0.1	inland waters	India
0.2	surface waters	Italy
0.2	mining effluents	Canada
0.2	bays and estuaries	Australia
0.3	sewers	Italy
0.5	surface waters	Argentina
1.0	sewers/marine waters	India
0.1 - 0.5	varies according	Denmark
0.1 - 0.8	to type/location of	Japan
0.3 - 2.0	plant and nature of	Germany
0.05 - 3.0	receiving water	FR
0.3 - 3.0		Belgium
		UK

(Source: International Lead and Zinc Study Group, 'Environmental and Health Controls on Lead' (1989))

Within the industrial environment there are two major areas of concern in respect of environmental controls:

a) Protection of the workforce in the lead industry against any significant exposure to lead or lead compounds. In the case of lead-acid battery manufacture, which is by far the greatest use of lead, the necessary environmental control measures are dealt with in Chapter 3.

b) Protection of the general public and lead recovery workers at the stage where the battery has finished its useful life, and has to be scrapped. This is dealt with in the following sections.

To illustrate the importance attached to these physiological effects of lead, Tables 4.5 to 4.9 show the upper concentration limits for blood lead, lead-in-air for workers in the lead industry, and airborne lead emission limits to the general atmosphere, which are internationally accepted.

4.2.3 Recycling of Lead-Acid Batteries

In view of the potential hazards of lead outlined above, it is fortunate that lead-acid batteries are one of the most eminently re-cycleable products of modern times, and also by far the greatest user of lead (Figure 4.3). In recent years, strong measures have been taken to eliminate lead from paints, and to remove lead pipes from plumbing. Currently, the encouragement to use lead-free petrol in road vehicles has high priority.

In Europe, a 1987 study (Bevington, 4.1.) estimated car battery recovery rates at 80 to 85%, although it is now believed to be considerably higher. In the USA, thorough calculations, (BCI, 4.2, 4.3), carried out by the battery industry's representative organisation, the Battery Council International, (BCI), for 1987 to 1990 showed that by 1990 the recovery rate had reached 97.8%. It is believed that for industrial batteries, the recovery rate approaches 100%, because of the volumes (hence value) involved, and because of the nature of the supply/return system operated.

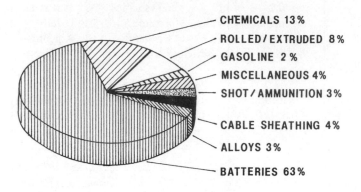

CHEMICALS 13%
ROLLED/EXTRUDED 8%
GASOLINE 2%
MISCELLANEOUS 4%
SHOT/AMMUNITION 3%
CABLE SHEATHING 4%
ALLOYS 3%
BATTERIES 63%

Fig.4.3. Breakdown of lead usage 1990

Despite these high rates of recovery, there have been periods when poorer rates have been experienced, notably at times of low lead price, when the economics of collection and transport simply become non-viable for the scrap collectors. Traditionally, the collection of spent SLI batteries has not been a particularly well-organised activity. However, increasingly severe environmental requirements in recent years have affected the economics of recovery in some areas and have resulted in a greater degree of involvement by battery manufacturers when the scrap trade alone has not been able to maintain the very high degree of recovery described above.

In consequence there are today two main routes for the recovery of scrap lead-acid batteries: the battery manufacturer who takes responsibility for his own product and organises collection through his retail outlets, and the scrap dealer who seeks out scrap batteries from all available sources.

Most of the methods that have been employed to try to improve the recovery rate depend on voluntary actions, and fall into two main categories: either education on the importance of recovery, or some financial incentive to return old batteries. Education about the benefits of recycling, such as economic, environmental, resource conservation, can encourage owners of spent batteries to take the trouble to return them. However, economic measures, such as part-exchange of the old battery for the new, are more effective. Indeed, in some countries, it is now a legal requirement to hand in an old battery when purchasing a new one, but this tends to be unduly bureaucratic, so it is not widely employed.

A number of other measures have been tried, including the imposition of levies, or taxes, on the sale of new batteries, with the funds so raised helping to promote better collection and return. Other schemes have used the incentive of a returnable cash deposit on the purchase of a new battery, but again this is rather bureaucratic. A new idea, recently proposed, is for the encouragement of recycling by requiring that a specified (and increasing) proportion of secondary lead is used in the manufacture of new batteries. This puts the onus on the battery manufacturers to ensure that the recovery methods work adequately. At the moment, this is only a proposal and has not yet been applied. Finally, another method is to apply legislation.

4.2.4 Environmental Legislation
In March 1991, a Directive (91/157/EEC) was issued by the Council of the European Economic Community (the EEC) on batteries and accumulators which contain dangerous substances (Council of the European Communities, 4.4). This directive is designed to ensure that all Member States take steps to ensure the collection and disposal of a variety of batteries, including lead-acid types. As mentioned earlier in this chapter, it is fortunate that the lead-acid system is particularly well suited to recycling, because a number of other battery systems which for example contain very toxic elements such as mercury, nickel or cadmium, are not so conveniently recycleable, so that their disposal presents even more onerous problems.

The EEC directive does not lay down many specific technical requirements (except, for example, a strict limit on the mercury content of alkaline manganese batteries), but rather concentrates on the standardised marking of types of batteries and the stipulation that Member States must set up ongoing programmes for the collection and safe disposal of batteries, with the details left to each State.

Thus, lead-acid batteries must be marked to indicate that they can be recycled, and that they must be disposed of separately, that is, not with household waste. The marking must include an internationally recognised recycling symbol which comprises three arrows in a ring. Consumers must be informed about the meaning of the markings and the dangers of uncontrolled disposal of the spent battery. The first programmes cover a four year period commencing on 18th March 1993, and subsequent updating will be carried out every four years.

An increasing number of countries are promoting schemes to improve recycling measures of lead-acid batteries, both in Europe and North America.

Sweden - Started formally in January 1991. The scheme is based on an environmental levy on all new SLI batteries (original equipment and aftermarket), which is administered by a Government agency, but which is collected and distributed by a joint company, Returbatt AB, whose members come from the battery industry, scrap collectors and secondary lead smelters. A 95% collection rate is the target.

British Columbia - Started formally June 1991. The scheme is based on an environmental levy on all new lead-acid batteries (SLI and traction), which is administered by a Government agency. Certain registered participants can apply directly to the agency for financial support in the form of a transport incentive. The target recovery figure is 98%.

Italy - Started formally September 1991. The Italian scheme is based on an environmental levy on all new lead-acid batteries (principally SLI and traction) which is administered by a joint company, COBAT, a non-profit making consortium with members from the battery industry, producers, garages, scrap merchants and Government.

USA. In the USA there is no Federal legislation designed to promote collection and recycling of batteries throughout the country. However, when increasingly strict Federal legislation on general environmental matters tended to force scrap dealers and other collectors to cut back on handling of scrap batteries, the battery industry stepped in with a voluntary scheme through its trade association, the BCI. The main planks of this scheme are to ban any land-disposal of scrap batteries, requiring all spent batteries to be handed in to an authorised collector, and to encourage collection of these batteries by requiring all retailers of new batteries to accept old ones in any numbers which are offered. Retailers are also required to display notices to this effect.

Other Countries. Battery collection schemes are receiving a great deal of attention in other countries, especially in the EEC States, where the directive described above, and which became effective in 1993, is causing urgent measures to be applied. A single pan-European scheme might have advantages, but is not likely to develop in the short term. So far, the experimental schemes in various countries have shown encouraging trends of improvement in the collection rates, and as these schemes become more established, it is expected that near 100% recycling rates will prevail for all major applications of lead-acid batteries.

4.2.5 Recycling Technology and its Impact on Battery Design
Once the difficult part of the recycling process has been achieved, namely the collection and transport to the treatment centre, it is obviously necessary to ensure that the efficiency of the physical and metallurgical processes used to reclaim all the battery components is as high as possible. As stated above, the modern lead-acid battery, and most of its components, are eminently recoverable, irrespective of whether they are of an automotive, traction or standby design.

The battery reclaim process benefits from the traditional lead metallurgy which has been developed over hundreds of years for the primary smelting of lead from mined lead ores, and in more recent times, the secondary (recycling) smelting of lead. Secondary smelting accounts currently for about 63% of total lead production in Europe, and the proportion continues to rise steadily.

The metallic lead components from a battery breaker need only to be subjected to a refining process, which is much less complicated than a smelting process. This latter is required when lead oxides, sulphates and so forth are involved.

The same relatively easy refining process applies to scrap lead in the form of old lead piping (not much used now), lead sheet roofing, and cable sheathing from old power and telephone cables, all of which can be relatively easily refined.

In secondary smelters, scrap batteries are crushed and smelted. The polypropylene from the containers (which now forms the majority of the container material used) is recycled to produce secondary plastic for automotive components, garden equipment and a range of other uses. The decreasing number of hard-rubber containers in use can also be crushed and find application as a cheap filler in horticultural containers, or in some operations, is used as a reductant in the smelting process. The lead metal is readily recoverable and re-used in battery manufacture. The lead compounds, largely oxides and sulphates, are converted into a suitable form to be added to the re-smelting cycle or converted to usable sulphates, depending on their chemical state. The sulphuric acid is also chiefly converted into usable sulphate compounds.

The only parts of the lead-acid battery which cannot be recovered are the separators or plate envelopes, whichever are used, and occasionally, small plastic components added as polarity markers or for other functional reasons. These consist of plastics such as polyvinylchloride, polyethylene, or acrylonitrile/butadiene/styrene, which is only just beginning to be effectively separated. However, these less-reclaimable parts of the battery form less than 5% by weight, and manufacturers continue to try to reduce their use.

4.2.6 Process Description

a) *Battery Delivery.* The incoming batteries are inspected to exclude any foreign material, and then tipped from a considerable height onto a hard inclined base, which cracks the containers and allows much of the sulphuric acid to drain away into a sump, from which it is pumped away for neutralisation. This is done by adding liquid sodium hydroxide and carefully monitoring the pH value.

$$H_2SO_4 + 2NaOH = Na_2SO_4 + 2H_2O$$

b) *Battery Breaking.* The cracked batteries are now tipped by a form of mechanical digger into a feed hopper, which feeds them into a battery crusher or hammer mill. This process reduces the batteries to small pieces, with a maximum size of about 5cm

The mixture is then fed through a large mesh sieve, onto a sloping vibrating screen under water sprays. The paste from the plates (lead oxides and lead sulphate) is washed out as a slurry, and is taken down into a collection tank, where it is further treated (see section c). The metallic lead, comprising much of the negative plates, the grids, pole pieces, connecting straps etc., along with the polypropylene fragments, pvc/paper separators and hard-rubber fragments, come off the bottom of the vibrating screen and are conveyored away to the hydrodynamic separation process (section d). See Figure 4.4.

c) *Paste Desulphurisation.* The paste slurry from b) above, containing lead oxide, lead

110

dioxide and lead sulphate, is concentrated by decantation, and treated with either sodium hydroxide or sodium carbonate.

$$PbSO_4 + 2NaOH = Na_2SO_4 + PbO + H_2O$$

$$PbSO_4 + Na_2CO_3 = Na_2SO_4 + Pb\ CO_3$$

The lead oxide or lead carbonate, depending on which method is used, is kept in suspension by stirring, and is pumped away to a filter press, where it is pressed into a cake for transition to the smelting process (see Section f). The filter cake is thoroughly washed to remove any remaining sodium sulphate. The sodium sulphate from this process is combined with that from the delivery bay (see Section a), and recrystallised to give anhydrous sodium sulphate of detergent grade for resale to detergent manufacturers or glass-works. See Figure 4.5.

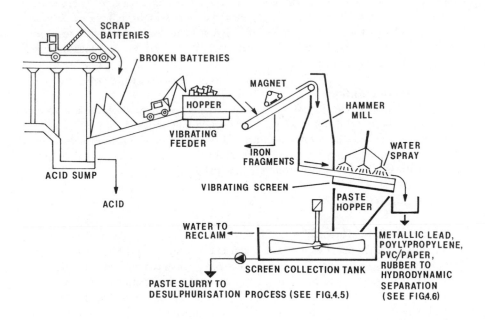

Fig.4.4. Battery breaking and crushing

d) *Hydrodynamic Separation Process.* The solid fragments coming off the vibratory screen, as described in section b) above, are fed into a special water tank with a water inflow at the base, and an overflow at the top. The metallic lead pieces sink to the bottom and are extracted on a rising inclined conveyor, whilst the polypropylene, hard-rubber and separator pieces float to the surface. Here, the polypropylene fragments are separated, washed and taken off for resale, whilst the hard-rubber and pvc/paper pieces are taken off to a static flotation tank, where the hard-rubber pieces are allowed to sink to the bottom, whilst the paper pieces float to the top. The hard-rubber pieces can be re-used as reductants in the smelting process because of their high carbon content and the high cost of metallurgical coke. The pvc/paper separators are not re-usable, and have to be dumped. See Figure 4.6.

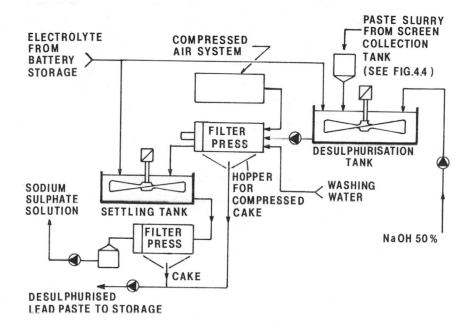

Fig.4.5. Paste desulphurisation

c) *Effluent Treatment Plant.* Because of the very considerable quantities of process and washing water used in the operations described above, it is clearly necessary to have a good water recovery process. The effluents from each of the processes are combined and the pH carefully adjusted. The combined effluent is then pumped through a series of fabric membranes with decreasing filter porosity so that its purity can be controlled to meet the appropriate legislation standards. Some of the water will have been recycled, but the final effluent is pure enough to be discharged to the local sewage system by agreement with, and after checking by, the local authority. The maximum allowable effluent discharge limits at present in force in the UK are as shown in Table 4.10.

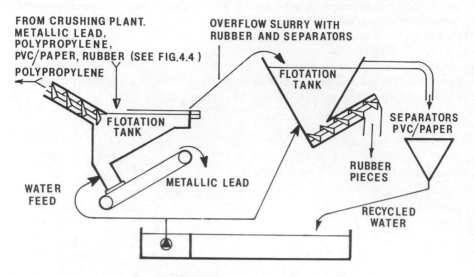

Fig.4.6. Hydrodynamic separation process

Table 4.10. Maximum allowable effluent discharge limits

pH value	6.0 to 9.0
Volume of discharge (m³/day)	200
Biochemical Oxygen Demand	10mg/1 in 5 days at 20°C
Dissolved antimony (mg/l)	5
Dissolved lead (mg/l)	1
Dissolved zinc (mg/l)	1
Dissolved arsenic (mg/l)	0.2
Dissolved silver (mg/l)	0.1
Dissolved copper (mg/l)	0.1
Dissolved nickel (mg/l)	0.1
Dissolved cadmium (mg/l)	0.05
Dissolved mercury (mg/l)	0.05

f) *Smelting of Lead Paste.* The standard equipment of the secondary lead industry used to be the blast furnace. However, in Europe, this has largely gone out of favour because of the high price of metallurgical coke, and the relative difficulty of preventing the escape of dust and fumes. The blast furnace provided a low grade antimonial lead which was converted to a purer lead in a refining kettle or a reverbatory furnace. The high-antimony slags were accumulated for a subsequent blast furnace treatment to produce a high-antimony bullion, which was appropriate for blending the high-antimony grid alloys which were used in earlier days.

Most companies now use rotary furnaces, either oil or gas fired. The charge can be tailored to give a lead of the approximate composition desired, or a two-stage smelting procedure can be used.

Stage One. Battery plates are charged into the furnace using little or no reducing agent, and crude soft lead is tapped off after a few hours, leaving the antimonial slag and some lead oxide/sulphate in the furnace.

Stage Two. Coke or anthracite fines and soda ash are now charged into the furnace, and both lead and antimony oxides and lead sulphate are reduced, and the cycle ends with the furnace being emptied of antimonial lead, and of slag for discarding.

Some companies use hard-rubber battery cases as co-reductants because of their high carbon content, and this reduces the need for expensive metallurgical coke, but the proportion of hard-rubber cases compared to polypropylene cases is steadily decreasing. Iron may be added to the charge in small amounts to compound with the sulphides produced from the reduction of sulphates, and to prevent any sulphurous fumes from escaping from the furnace.

The most modern secondary lead furnace is the Isasmelt furnace, so called because it was developed by Mount Isa Mines Ltd in Australia. It consists of a refractory-lined cylindrical vessel (Figure 4.7) of 1.8m internal diameter, and is capable of handling 35,000 tonnes of lead bullion per annum. Desulphurised paste from operation c) above along with reductant are continuously fed into a bath of molten lead and lead components. The submerged combustion lance, through which high-pressure air and oil are pumped, agitates the bath, and produces a soft bullion with a lead content of 99.9% that is intermittently tapped off into a storage pot system.

At the end of each cycle, the furnace slag is treated and reduced to produce a high-antimony bullion and a discard slag. The reduction stage to produce the final slag is then commenced. Coal is added to the charge, and the oil flow in the lance is increased to raise the temperature in the furnace to between 1150 and 1250°C. A ferrosilicate slag with a low lead content is then drained from the furnace, along with any metal remnants.

The Isasmelt process is claimed to address environmental concerns and give reduced operating costs. Its main advantages over the traditional rotary furnace are stated to be:
- high thermal efficiency and low operating costs,
- the elimination of soda fluxes,
- direct production of both soft lead and antimonial lead alloy giving alloying flexibility,
- ability to produce low-lead discard slags, facilitating the ease of environmentally acceptable disposal,
- good process-hygiene due to the semi-continuous nature of the process.

114

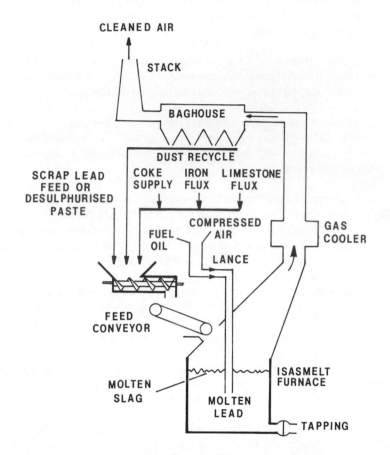

Fig.4.7. Schematic diagram of the Isasmelt furnace

g) *Secondary Lead Refining.* Once the smelting process is completed, the molten lead is removed from the furnace and pumped into the refinery pots. The principal impurities which are removed are copper, tin, antimony and arsenic. Copper is removed in similar fashion to that described above for primary lead. Traces of bismuth and silver are usually present, but these are not normally removed since traces of silver are found to improve the corrosion-resistance of grid-lead alloys, and the small amount of bismuth is not harmful.

h) *Gaseous Effluent Cleaning.* As with primary smelting, large volumes of gas are produced in this secondary smelting procedure, carrying with them substantial quantities of dust. On leaving the furnace, the gas is cooled from about 900°C to about 100°C by air- or water-cooled heat exchangers. The gases pass into a baghouse which contains hundreds of woven cloth bags. The gases pass through the bags in series/parallel, and the dust remains on the inside surface of the bags. Periodically, there is a negative back-pressure, and the airflow to a group of bags is cut. The dust cake cracks and falls to the floor of the bag chamber, from where it is extracted and returned to the smelter. The cleaned gases pass out of the stack and into the atmosphere, dust-free. In the course of processing one tonne of lead through the smelter, as much as 100 tonnes of air have to be cleaned in this manner.

In Europe and the USA, which together account for 75% of world secondary lead smelting, many producers are making operating losses due to low lead prices. In addition, lead smelters are facing a rise in costs because of ever-rising environmental expenditure. Additional costs could force the closure of plants producing up to 200,000 tonnes per annum in the USA. A similar situation exists in Europe. The most vulnerable operations are small to medium sized smelters, owned by either independent companies or companies with no other lead interests. While smaller smelters face problems, large up-to-date secondary smelters are competitive. Upgrading larger operations and major expansions should enable competitiveness to be maintained.

i) *Plastic Reclamation.* The most modern plastic reclaim operation in Europe is probably the joint operation by Cookson-Pennaroya, in Villefranche, France. In 1991, it produced 10,000 tonnes of reclaimed polypropylene per annum from 50,000 battery containers per day, and a turnover of about £6m. The technology developed by the company enables the process to separate, reclaim and produce high-quality materials which are suitable for use by the automotive industry at a price competitive with virgin material.

Each day, the production lines treat more than 40 tonnes of polypropylene scrap from the crushing of 50,000 batteries. This means that every year, 40,000 cubic metres of battery scrap are saved from dumping, so that the benefit in environmental terms is considerable. The process employs fully controlled effluent treatment and closed-circuit water cooling systems. It sorts valuable polypropylene from less valuable materials such as acrylonitrile/butadiene/styrene, polyvinylchloride, bakelite, ebonite, polyurethane and rubber. It employs various stages of crushing, hydrodynamic separation, centrifuging and pneumatic separation by warm air. Sorted polypropylene particles are plasticised and mixed with additives for stabilisation and colouration to produce pellets ready for re-use.

Because recycled products are not yet well accepted, the market strategy is aimed at three major markets:

1) The automotive industry. The automotive industry is most advanced in using reclaim polypropylene for a variety of uses, most notably for the moulding of inner wheel arches and bumpers.
2) Horticultural applications. With the advent of reactive extrusion which allows modification of the fluidity of the polypropylene, materials with

high melt indices can now be obtained. These are especially suitable for thin-walled mouldings such as horticultural containers, in a range of non-pastel colours.

3) Technical applications. The process also allows for the production of thermo-plastic rubber components, which are produced in various grades with the addition of talc, carbonate or glass-fibre filled additives with a wide range of melt indices. These are useful for products such as electrical equipment boxes, switchgear housings and many other rigid containers.

4.2.7 Conclusion. Environmental Acceptability of Lead-Acid Batteries

As environmental concerns grow, and as environmental legislation tends to become stricter in many parts of the world, it is fortunate that the lead-acid battery in its many applications is well suited to near 100% recycling. High recovery rates are somewhat susceptible to fluctuating market prices, which is a factor outside the control of the lead industry, and consequently, the industry has to work very hard, both to keep up collection rates of scrap batteries, and to improve processing efficiency of the recovery technology. However, here again the lead-acid battery is well placed against its competitors, virtually all of which either have a more difficult recycling process involved, or, if disposable, represent a considerable turnover of potentially toxic chemical substances to waste.

REFERENCES

4.1 C.F.P.Bevington (Metra Consulting Group) 'Lead-Acid Battery Wastes in the EEC', Report prepared for the Commission of the European Communities Directorate General X1 on Environment, Consumer Protection and Nuclear Safety,1987.

4.2 Battery Council International 'National Recycling Rate Study' Chicago, Illinois, 1991.

4.3 Battery Council International '1990 National Recycling Study' Chicago, Illinois, 1992.

4.4 Council of the European Communities. 'Council Directive of 18 March 1991 on batteries and accumulators containing certain dangerous substances' (91/157/EEC), Official Journal of the European Communities, No. L78/38, 26 March 1991.

CHAPTER 5

Battery Operating Characteristics and Behaviour

As noted in Chapter 2, the modern automotive battery is specifically designed to provide high current rates for engine cranking, coupled with a comparatively small amount of reserve capacity, to cope with the quiescent loads and emergency functions when the charging circuit fails, at the minimum weight, volume, and cost, commensurate with the other physical functional requirements of the application.

The main feature of the design is the spreading of the solid active materials across as great an area as is practical, to make available the highest current possible. This is achieved by maximising the pore structure of these materials, and spreading them very thinly over as great a grid area as can be practically designed into the volume available. For any given current, therefore, the current density, i.e., the current per unit of area, often expressed as amps/sq.cm or mA/sq.mm, is as low as can be achieved. Given that the voltage across the electrodes is inversely dependent on current density, it follows that such a design provides the best battery voltage for any given current.

5.1 Rating

The automotive battery 'rating', i.e., its performance specification, is normally given in terms of a high-rate current, normally in the range of 100 to 500 amps, and a time in minutes, of the order of 20 to 60 minutes, reflecting the reserve capacity available (see Section 2.11). These numbers enable comparison to be made against an application, with other similar automotive batteries in this comparatively narrow field of function, that is, at high rates of discharge.

5.2 Capacity

The battery has capability outside high-rate discharge, and batteries are often 'rated' in terms of ampere-hour capacity, measured at a much lower constant discharge current, to a predetermined cut-off voltage.

Ampere-hour Capacity = Amperes of Discharge x Time in Hours

The traditional practical standard is the current that will discharge the battery to a voltage of 1.75 volts per cell, at 25°C, in 20 hours. This is known as the 20 hr rate, and the capacity obtained is the 20 hr capacity, or C_{20}. Typical automotive batteries for private car use would have a C_{20} in the range of 40-60 Ah.

The 20 hr capacity is normally within a few percentage points of the maximum practical capacity obtainable from a battery of this type.

The lower the discharge current the greater the ultimate conversion of lead dioxide and lead into lead sulphate, assuming an adequate quantity of sulphuric acid is present. However, as discussed elsewhere, the structure of the active material on the plates is such that only a limited amount of conversion is possible, and in practice the best that can be achieved without sacrificing a useful life is of the order of 50% of the theoretical limit of the masses of lead dioxide and lead present in the active material, calculated using Faraday's Laws.

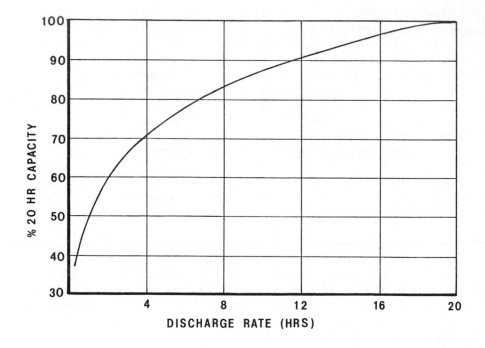

Fig.5.1. Capacity against discharge rate for a typical lead-acid automotive battery at 25°C

Figure 5.1 demonstrates that capacity falls away as discharge current is increased. This is due to a number of factors. Sulphuric acid is needed as part of the reactions, and as the reaction rate increases, i.e., producing a higher current, so the rate of acid diffusion deeper into the pores, as time progresses, becomes more and more of a limiting factor. The product of the reactions, lead sulphate on both of the plates, has a very low level of solubility in water, and in sulphuric acid solutions; a thicker and thicker layer is produced at the interface with the reactants, decreasing the material available for the reaction, and increasing the solid volume within the pores, further decreasing acid migration. As these effects happen, so the voltage of the battery drops, due to an increase in the internal resistance.

A small increase in capacity can be obtained by discharging at rates lower than I_{20}, the 20 hr current, but the curve becomes asymptotic to the x-axis, as Figure 5.1 suggests, and the gain is small.

Many attempts have been made to predict the behaviour of the lead-acid battery under a variety of conditions, by developing equations relating the effects of these processes. Some of these will be examined later in this chapter, but a fairly simple and practical equation relating discharge rate to time of discharge, at a constant temperature, is that developed by Peukert, referred to in Chapter 2.

Peukert states that :

$$I^n t = C$$

where I is the discharge current, and t is the discharge time. n and C are constants for particular batteries, and can be evaluated by discharging a battery at two current rates, I_1 and I_2, producing discharge times of t_1 and t_2.

$$I_1^n t_1 = C$$

$$I_2^n t_2 = C$$

$$n \log I_1 = \log C - \log t_1$$

$$n \log I_2 = \log C - \log t_2$$

$$n = \log t_2 - \log t_1 / \log I_1 - \log I_2$$

Once n and C are evaluated the equation can be used to predict capacity at different rates of discharge. The equation can be used quite reliably over a large range of discharge rates, but becomes less accurate at very high rates of discharge, i.e., when discharge times are of the order of minutes or less.

Battery capacity is temperature-dependent, as shown in Figure 5.2. The chemical reactions responsible for current take place at a slower rate as temperature reduces. In addition the viscosity of the electrolyte increases as shown in Figure 5.3, and resistivity also increases (see Figure 5.4).

As an approximation, at low rates of discharge, battery capacity reduces by

120

1% of the 20 hr capacity for every 1°C reduction in temperature. However, as the rate of discharge increases so the percentage of available capacity decreases.

The other factor affecting battery capacity is the sulphuric acid electrolyte concentration (and total quantity). The viscosity and conductivity of the electrolyte both change as the concentration of acid changes, as can also be seen in Figures 5.3 and 5.4. Electrode potential is also a function of electrolyte concentration. Clearly, as sulphuric acid is one of the reactants, the total quantity available can have an effect on capacity.

The concept of **Watt Hour Capacity** can be useful in comparing batteries designed for different purposes.This is a measure of the total energy available, or the time for which a given level of work can be carried out. As the voltage of a lead-acid battery reduces during discharge, for reasons given below, the average voltage must be used in calculating watt hr capacity

Watt Hr Capacity = Amp Hr Capacity x Average Voltage

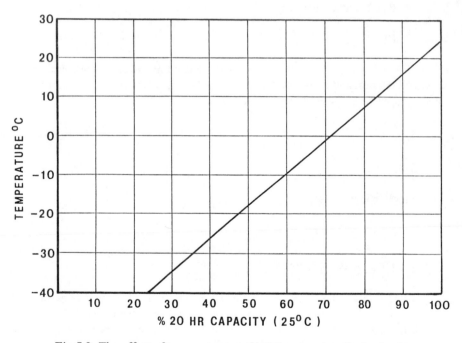

Fig.5.2. The effect of temperature on the 20hr capacity of a lead-acid automotive battery

The modern automotive battery can provide up to 45 watt hrs per kilogram of total battery weight, compared with a figure of 20 to 30 watt hrs per kilogram for a typical industrial battery. This is offset by the large number of complete charge/discharge cycles that the industrial battery can provide, which the automotive battery cannot. The automotive battery spends most of its time in a nearly fully charged condition. The capacity removed by starting the vehicle is small, and is quickly returned by the charging system.

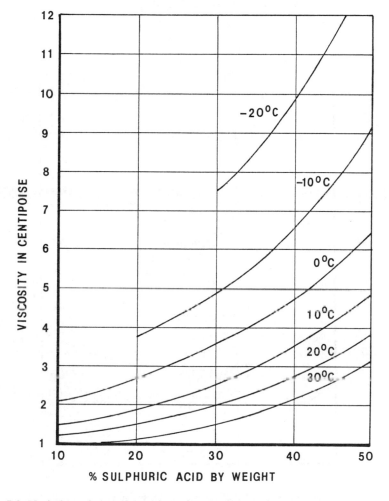

Fig.5.3. Variation of viscosity with acid concentration at different temperatures for sulphuric acid solutions

122

5.3 Open-Circuit Voltage

The open-circuit voltage, or, more correctly, the Electromotive Force (EMF), is the voltage across the terminals of a fully charged cell or battery at rest, i.e., with no current flowing. In the case of a lead-acid cell at 25ºC and with a sulphuric acid electrolyte of 1.280 specific gravity, this is normally given as 2.125 volts per cell, or 12.75 volts for a standard 6 cell battery. This value varies a little with temperature, not significant in practice, and varies with acid concentration as shown in Figure 5.5.

It should be noted that, when measuring the open-circuit voltage of a battery after charging, some time should be allowed before taking the measurement, to permit the diffusion processes, etc., to reach equilibrium.

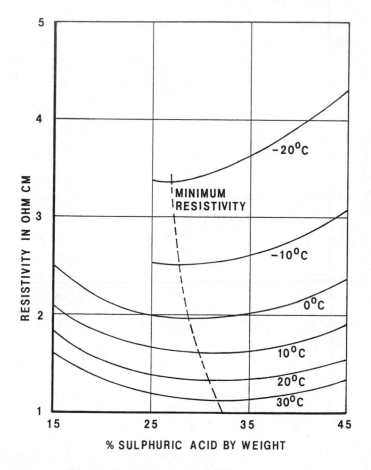

Fig.5.4. Resistivity of sulphuric acid solutions

5.4 Discharge

5.4.1. Discharge under Load

Immediately a load is placed across the terminals of a battery, and current flows, the voltage falls, due in part to the ohmic resistance of the battery, and in part to polarisation effects.

At low rates of discharge, this initial drop from the open-circuit voltage is quite small. Figure 5.6 shows discharge curves for a battery at low constant current rates, i.e., for a typical 40Ah battery, (20hr rate), the 40hr rate is approximately 1A, the 20hr rate is 2A, and the 10hr rate is a little less than 4A. At higher rates, the departure from the open-circuit voltage increases. Voltage continues to decline as discharge progresses, as the rate at which sulphate ions diffuse into the pores of the active material becomes more and more inadequate to sustain the constant current, and as the bulk conductivity of the electrolyte declines from its maximum at about 1.225 specific gravity. Towards the end of discharge the drop in voltage becomes more distinct, and faster, reaching a point where the battery is virtually exhausted. The cut-off voltage used in standard tests is this point, where the current/time discharge curve forms a 'knee' and there is little practical use in continuing the discharge.

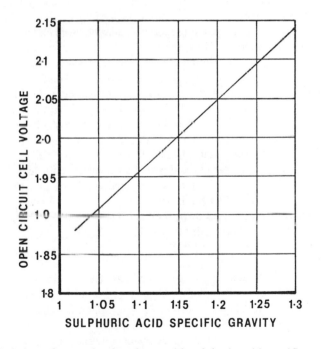

Fig.5.5. Variation of open-circuit voltage with sulphuric acid specific gravity for the lead-acid system at 25°C

This point becomes lower, and less pronounced, the higher the discharge current. Acid concentration falls linearly with discharge time at constant current, in accordance with Faraday's second law, and, provided that adequate time is allowed, to enable acid diffusion to produce a homogeneous solution, measurement of specific gravity provides a quite accurate means of checking the state-of-charge of a battery.

At very high rates of discharge and low temperatures, as Figure 5.7 shows, discharge voltage is depressed very significantly, and discharge time is very much reduced. This is just the circumstance simulating engine starting on cold winter mornings, and it is clear from the figures that lengthy engine cranking due to difficulty in starting can quickly exhaust even an initially fully charged battery. Figure 5.8 indicates the approximate capacity available at these high currents and low temperatures. To complete the picture, Figure 5.9 gives an indication of the effect of temperature on capacity at high rates of discharge.

5.4.2 Self-Discharge

The lead-acid battery can slowly lose charge when left idle. Different chemical materials have different electrochemical potentials, and when two are in contact with an electrolyte, such as sulphuric acid, an EMF can exist between them, and if a conducting path is also present a current can flow. Hence, on the plates of a lead-acid battery, small localised cells can be formed between the active material that should be present and any small inclusion of an impurity, leading to the discharge of that element of the active material. During regular use, and therefore charge, of the battery on a vehicle this goes unnoticed. However, it can be noticed in batteries that are only infrequently or irregularly used. Until quite recently, for instance, where SLI batteries were used for seasonal applications, such as caravan or small boat use, it was frequently recommended that batteries left unused for the winter should be recharged at least once a month, to prevent discharge, and hence sulphation, from occurring. However, modern technology has produced purer components which suffer much less, and for such applications a charge prior to storage for the winter should be sufficient. Indeed, monthly charging can damage the battery and reduce its life.

A further point to bear in mind is that an electrical path across the external top of the battery can exist if the top is dirty and traces of moisture containing sulphuric acid or other conducting fluids are absorbed. Such a situation can also give rise to a loss of charge during periods of non-use. The answer is to keep the external surface of the battery in a clean and dry condition.

5.5 Charging

Because of the reversible nature of the reactions involved, charging can be carried out by providing a direct current through the battery in the reverse direction to that of the current drawn when discharging. This can be done by applying a reverse voltage somewhat greater than the natural voltage of the battery.

Charging may be carried out using a constant current with a gradually increasing voltage, by applying a constant voltage with a gradually-reducing current, or by a combination of the two conditions.

5.5.1 Constant Current Charging

Provided a sufficiently high voltage is available as a supply, a constant current recharge can be applied to the battery, with a variable resistance in the circuit to constantly correct the current as the back EMF of the battery increases with charge.

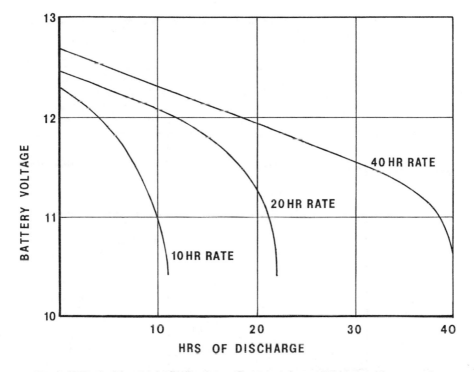

Fig.5.6. Typical low-rate discharge performance for a 12V lead-acid automotive battery at 25°C

As can be seen from Figure 5.10, in this instance where the applied current is $0.1C_{20}$, i.e., 5 amps for a battery with a capacity of 50 Ah at the 20hr rate, the voltage gradually increases with charge as the lead sulphate is reconverted to lead dioxide at the positive plate, and to lead at the negative plate, and the sulphate ions are released back into solution. However, as the battery approaches full charge, the increase in voltage accelerates as the lead sulphate available for conversion is reduced, and a smaller percentage of the current is used for this purpose. In order to sustain the current, the potential on the electrodes increases to a level where other electrochemical reactions can take place. These include :

$$4OH^- - 4e = O_2 + 2H_2O$$

at the positive plate, and

$$2H^+ + 2e = H_2$$

at the negative plate, i.e., oxygen gas is evolved at the positive plate, and hydrogen gas is evolved at the negative plate.

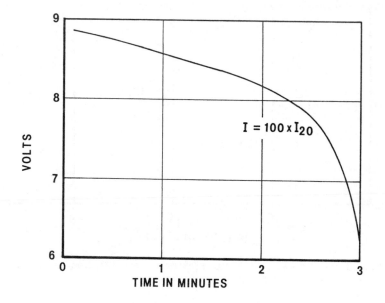

Fig.5.7. High-rate discharge curve for a lead-acid automotive battery at -18°C.

Gassing need not start on both plates at the same time, or in equal amounts, as the relative amounts of material may differ on the two plates, and the potentials for these reactions may not be reached at the same time, but as full charge is approached so the proportions of the two gases become stoichiometric for the electrolysis of water, i.e., two molecules of hydrogen to one molecule of oxygen.

At low rates of charge, such as those shown in Figure 5.10, and relatively high temperatures, most of the current is used in re-charge, and the amount of gassing is small until full-charge is reached, at which point most of the current is used in gas generation. Full charge is indicated when three one-hourly readings of specific gravity remain constant, and all cells are gassing freely. However, the onset of gas evolution happens earlier in the charge as current is increased (the imposed voltage is higher to drive the higher current), and charging becomes less efficient. The effect can be mitigated somewhat by using grid alloys containing less antimony. Antimony has the effect of depressing the gassing potential.

5.5.2 Constant Voltage Charging

In this case a constant voltage is applied to the battery, normally with a current-limiting device built into the circuit, to prevent excessive early currents being passed. The voltage chosen needs to be at an appropriate level to minimise gassing on either electrode. In the case of modern electrical systems on vehicles this voltage is set at about 14.2-14.4 volts. The fully discharged battery will accept a very high current at this level, slowly reducing as charging proceeds. As the cut-off voltage is approached, and the battery approaches full charge, the current tails off to a very low level. Under such circumstances the battery would never reach a fully charged situation as the current would become infinitesimally small before 100% charge is reached.

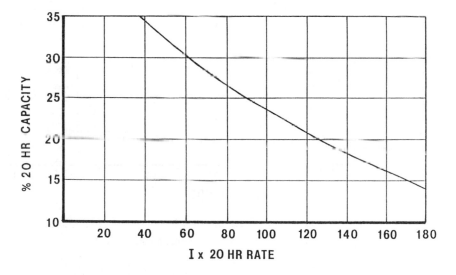

Fig.5.8. Variation of high-rate capacity with current at -18ºC for a typical lead-acid automotive battery

128

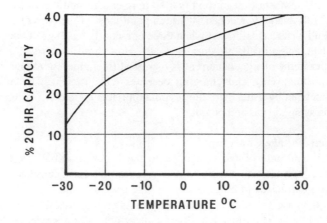

Fig.5.9. Typical high-rate performance against temperature for a lead-acid battery
(current at 100 x 20hr current)

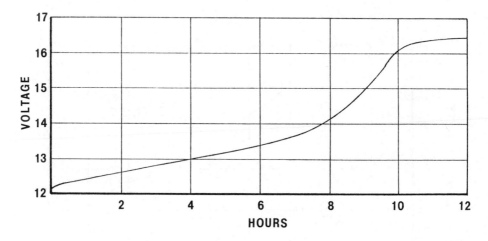

Fig.5.10. Typical constant-current recharge at $0.1C_{20}$ at 25°C for a lead-acid
automotive battery with antimonial-lead positive grids

In practice, on a vehicle, this is not important, as the running time, and
hence re-charge time, is normally more than adequate to replace the charge taken out

by starting, and the battery spends much of its working life 'floating' in a greater than 90% charged condition.

Off the vehicle, where charge needs to be put back in as quickly as possible, fast or boost chargers are normally of the constant voltage design. However, if a fully charged battery is required, for instance as a precursor to carrying out a capacity test, a constant current charger, often with a current step-down sequence, dependent on time or voltage, is used.

Before leaving the subject of charging it would be as well to point out some of the practical aspects to watch out for in carrying out the procedure:

1) Whilst it is bad for the battery to charge with an inadequate volume of electrolyte present, i.e., when the top part of the plates is left uncovered, it is equally undesirable to top up the battery before charging. The reason for this is that the volume of the electrolyte increases with increasing charge. It is thus possible, at the end of charge, to have electrolyte overflowing if the battery were to be topped-up prior to recharging. In addition, unless it is known that the battery has lost sulphuric acid, topping up should only ever be done with distilled water. If sulphuric acid has been lost great care needs to be exercised in replacing only the exact quantity necessary to restore the correct fully charged specific gravity.

2) It is essential to make sure that there is good contact at the terminals of the battery before commencing charge, that the battery is connected in the right polarity for charging, and that the charging circuit is broken remotely from the battery at the end of charge. As explained above, hydrogen and oxygen can be evolved from a battery towards the end of charge. Under some circumstances of concentration these gases can explode if sparked. Clearly the immediate vicinity of the battery is where such concentrations are most likely to occur. Even on the vehicle, immediately after the battery has been on charge using the vehicle electrical system, there is the possibility of some gas being generated, and it is worth waiting for a few minutes, to allow gases to dissipate, before carrying out any work in the vicinity of the battery.

3) It is undesirable to charge a battery at too high a temperature, or too high a current, which causes the temperature to rise excessively. It is also undesirable to charge the battery at too low a temperature.

4) Make sure that in connecting a battery to a vehicle the battery is connected with the terminals the correct way around. Modern vehicle electrical components (particularly electronic circuits) can easily be irretrievably damaged by even momentary wrong connection.

5.5.3 Charging to offset Self-Discharge

As described above in Section 5.4.2, in a number of circumstances charging is required to offset the self-discharge of a battery, and this charging need is dealt with in that section.

5.6 Battery Life and Failure Modes

As discussed in earlier chapters modern technology has done much to improve battery life-expectancy in a number of ways: use of polypropylene for containers, improved battery sealing, more corrosion-resistant alloys, better casting technology, better assembly methods, better control of on-vehicle charging, to name only a few improvements. Indeed, through the 1970s and the early 1980s, average battery life in most of the developed regions of the world increased from about two years to over four years. (These figures do not apply to tropical climates, however, because, although helped by the use of lower-gravity electrolyte, normally in the range of 1.240-1.250, the operating temperatures are such that a significantly reduced life is experienced.) However, in recent times little further improvement has been noticed. (Indeed, in some areas there has been a small reduction in life). This is largely due to two significant effects. Firstly the commercial pressure on down-sizing the battery, while maintaining high-rate performance, has led to much lower paste-densities, a reduction in the spacing between the plates, and the use of the minimum amount of electrolyte. The down-sizing is not only due to economic pressure, but also the need to pare all aspects of vehicle weight, in order to minimise fuel consumption. The second reason has been the move to lower bonnet or hood heights, to improve the aerodynamic shape of the vehicle, whilst introducing more under-bonnet equipment, both of which require compaction of the engine and surrounding components, hence reducing air-flow, and increasing under-bonnet temperatures.

Accurate figures of battery life are difficult to obtain. Accelerated testing methods give very poor comparison with real-time experience, where there are so many variables such as climate, application, usage patterns, installation, together with other environmental factors. In order to cover these factors it is necessary to compile data from a large sample-number of batteries that have failed in service, and analyse the results statistically. The most up-to-date information is that obtained from the BCI survey presented at the 107th BCI Convention at Orlando, Florida on the 3rd May 1995, and shown in Figures 5.11, 5.12, and 5.13, based on a sample of over 3000 passenger car batteries scrapped from vehicles across the USA Figure 5.11 is compiled from life-data from centres across the USA and demonstrates clearly the effect of the average ambient temperature on the life of a lead-acid battery.

Mode of failure is assessed by cutting open individual failed batteries, and examining them in detail for defects. A great deal of experience is required to find the reason for failure and to assign a cause. In this survey the causes of failure have been grouped into six categories as indicated in Figures 5.12 and 5.13.

1) Serviceable. This is clearly not a reason for failure in the strict sense. No fault was found in these batteries, and it is unlikely that a fault was present, as evidence of malfunction of the battery would normally be present following a pre-existing fault. The likely reason for scrapping is that the battery was in a sufficiently discharged state to render it unable to start the vehicle, and the user incorrectly diagnosed the cause. There may have been a fault in the charging circuit, a short-circuit fault elsewhere in the vehicle electrical system, or even that the vehicle had been inadvertently left with an item of electrical equipment switched on and unnoticed.

Whatever the reason, all such surveys include a significant proportion of batteries which have been incorrectly diagnosed as faulty, and which have simply needed a re-charge. This is not suprising, however, given the likely severity of the

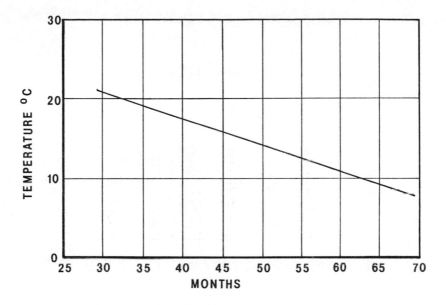

Fig.5.11. Average age to failure from manufacturing date

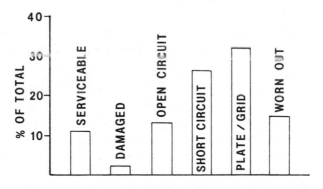

Fig 5.12. Reasons for failure as a percentge of total failures

breakdown, i.c., total immobilisation of the vehicle, often in trying circumstances, and the difficulty in accurately and quickly diagnosing such problems. Indeed, non-destructive testing of a battery to prove that it is fault-free, involving re-charge, capacity-checks, and loss-of-charge testing, can take several days.

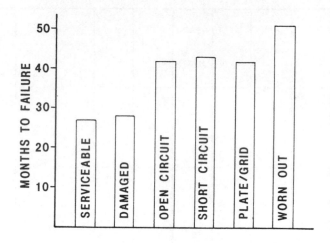

Fig. 5.13. Average time-to-failure of the reasons for failure

2) Damaged. This category tends to include all those faults often caused through removing or replacing the battery on the vehicle. Such faults include broken containers/lids caused by heavy blows or puncturing, terminals twisted off during disconnection, fused terminals caused by bad reconnection etc. One notable cause in this category is that due to explosion, caused either by a spark at a terminal or internally due to a broken cell-to-cell contact.

Categories 1) and 2) do not contain faults attributed to normal wear and tear, and tend to be random events during a battery's life. Hence, as indicated in Figure 5.13, these categories have the shortest average life.

3) Open-Circuit. Any loss of continuity inside the battery will cause the battery to fail with an open-circuit, i.e., zero volts between the terminals. This is normally due to a break at one of the lead-lead interfaces, either the cell-to-cell interconnection, the connection between the interconnector and the strap holding a group of plates within a cell, or the connection between a strap and one of the terminals. Severe vibration or mechanical shock can cause a crack at one of these positions which, perhaps after subsequent corrosion within the crack, can cause failure of the joint. Occasionally manufacturing defects, an inclusion in the metal, poor intercell welding, or simply inadequate material cross-section, can be the cause.

The three remaining categories are mainly associated with the chemical or electrochemical reactions that take place within the battery under a variety of circumstances.

4) Short-circuit. If, within a cell, an electronically-conducting path between the positive and negative halves of the cell develops, this will in part or, if large enough, wholly replace the ionic path within the cell, thus partly or wholly removing that cell from the normal battery function, evident by the reduced voltage available on discharge. This can occur in a number of ways. 'Bridging' with corrosion product, or particles eroded from the plates, can occur around the edges of opposing plates, if there is separator damage or insufficient separator overlap. This material can also build up on the top of plates, and 'bridge' to the opposing plate strap situated above if the gap is too small. Pinholes in separators can cause crystalline extensions to grow through such holes, and short on to the opposing plate. Finally, towards the end of normal life, sediment build-up in the bottom of the cell can cause shorting across the bottom of opposing plates.

5) Plate/Grid. In this category are the reasons for failure that tend to affect all the same types of component in the battery more or less equally, and are usually caused by usage patterns outside the norm, i.e., extreme temperatures, non-automotive service applications and so on. Occasionally, such faults can be caused by a poor, or incorrect, paste formulation at the manufacturing stage. Grid corrosion, poor paste adhesion, negative paste shrinkage and soft 'muddy' positive paste all give rise to a gradual loss of capacity and ultimately to an inability to start the engine. Grid corrosion can be due to inadequate manufacturing processes or inferior grid alloy, although these causes have been much reduced in recent years; or, more often, is now caused by too-frequent overcharging of the battery – this cause of failure is more properly dealt with under category 6 *(Worn out/Abused)*.
In the survey, this category *5)* also includes failure due to sulphation, caused normally by allowing the battery to stand in a discharged state for a long period, and failure due to loose plates caused by severe vibration or poor welding in the manufacturing process.

6) Worn Out/Abused. The battery will eventually wear out, due to active material being shed from the plates, loss of conducting contact between particles of the plate and from particle to grid metal, with a consequent loss of capacity, to a level where the battery will no longer be able to carry out its prime function, to start the vehicle. This is the normal 'wear out' situation and the loss of material integrity is due to a mixture of causes: vibration, perhaps some loosening of the active material during gassing and the continual cycling, i.e., expansion and contraction of particles of active material between the charged and discharged state.
In the survey, other failure modes are included in this category. Of

particular note is the failure due to low electrolyte levels. With some modern batteries, access to the battery for topping-up with water is limited or non-existent. This mode of failure will become more commonplace, as some water-loss will take place by natural breathing and there will be some gassing during the normal course of a battery's life. As referred to earlier in category *5)*, continuous or frequent overcharge can, of course, produce severe water-loss, positive grid-corrosion, and excessive shedding of positive active material. In general, many motorists and more particularly those people who use automotive-type batteries for leisure activities, such as caravanners and boating enthusiasts, tend to subject their batteries to considerable overcharge.

5.7 State of Charge Measurement

The concept of battery capacity was discussed in Section 5.2. This defines the charge that can be stored in a new and fully charged battery. However, it is often important to know the amount of charge remaining in a partially discharged battery and this is more difficult to determine accurately. A battery's 'state of charge' is usually defined as the charge available from the battery, expressed as a fraction of the rated capacity (rated capacity and the capacity variation with discharge rate are discussed earlier in this chapter). Thus a 50 Ah battery with 25 Ah available on discharge would have a state-of-charge of 50%. It is obvious that state of charge is dependent on the rated capacity and expected discharge rate, so automotive battery state of charge is most frequently expressed in terms of the 20h rate capacity for convenience.

State of charge can be a vital parameter in many applications. In automotive use, it is of vital importance to know if the battery is capable of starting the vehicle. Knowledge of the state of charge also allows the most appropriate recharge rate to be selected, and any change during vehicle operation can be used as a diagnostic tool for the health of both the battery and for the rest of the vehicle's electrical system.

There are two principal ways of determining state of charge for a lead-acid battery. These can be classified as:

a) Physical Methods - based on following chemical and physical changes taking place in the battery cell on discharge.

b) Electrical Methods - based on following the changes in electrical parameters of cells or complete batteries on discharge.

5.7.1 Physical Methods

As noted in Chapter 2 and earlier in this chapter, the change in acid concentration in the cell electrolyte is an unambiguous function of the state of charge. One significant by-product of this is the change in cell open-circuit voltage with state of charge and this is referred to later. However, a variety of methods have been used in the past to determine state of charge by measuring the acid concentration in the cell. The change in some acid solution properties in the concentration range of interest is shown in Figures 5.14a, 5.14b and 5.14c. These figures show the properties as a function of theoretical electrolyte capacity (based on the sulphuric acid required in the cell reaction).

Classically the change in specific gravity (Figure 5.14a) has been followed by means of various hydrometer arrangements. A specific gravity of 1.250 (at 25°C) indicated virtually full-charge, falling to about 1.100 on complete discharge. Many earlier motorists used a glass pipette containing a hydrometer float to sample free acid from battery cells. The float would be calibrated directly in approximate state of charge rather than specific gravity. In some batteries, different coloured floating balls (with appropriate density ranges) were used to indicate approximate specific gravity according to whether they sank or floated. In larger and more sophisticated batteries (but rarely in automotive types) hydrometer floats were equipped with electrical position sensing as an indication of state of charge.

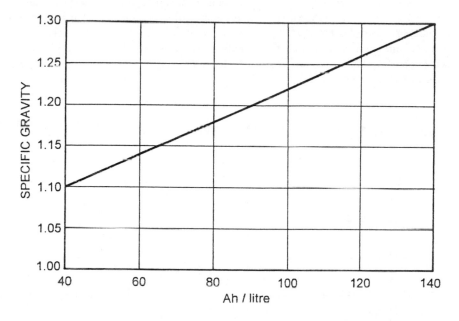

Fig.5.14a Specific gravity as a function of theoretical electrolyte capacity

In other cases refractive index (Figure 5.14b) was used as the concentration indicator. One popular device gives a change in colour of a 'cat's-eye' indicator built into one cell to assess state of charge. The availability of optically active semiconductor devices has also meant that a continuous electrical reading of refractive index can be obtained with simple fibre-optic refractometers (Hancke, 5.1).

It is unfortunate that conductivity shows a point of inflection within the operating range and cannot be used as an unambiguous indicator (Figure 5.14c).

136

Such electrolyte sensors suffer from a variety of interference effects. Dissolved gases in the electrolyte deposit small bubbles onto hydrometer floats or refractometer interfaces, giving significant errors unless removed by agitation. Also, ageing batteries deposit debris from electrode degradation onto the sensors, also giving inaccuracies. The reading from a single cell within a battery may not be representative of the battery as a whole and, in automotive applications at least, it is not economic to provide more than one sensor per battery. Even in a representative cell, the sensor can normally only be placed in electrolyte that is accessible at the sides, or at the top or bottom of the plates. There is an appreciable time-lag between the concentration of this electrolyte contained within the plate structure and that around the sides when rapid charges or discharges occur. Electrolyte at the bottom of cells may also become more concentrated than the average unless extra agitation occurs.

The above disadvantages have tended to discourage the widespread use of physical sensors in recent years, though work still continues for various battery applications. Besides the electrolyte variations described, there are also changes in plate volumes and in active material conductivities which can be followed if the right sensors are available (Calabek et al, 5.2). Reference electrodes may also be used to indicate electrolyte concentration (Tsubota, 5.3).

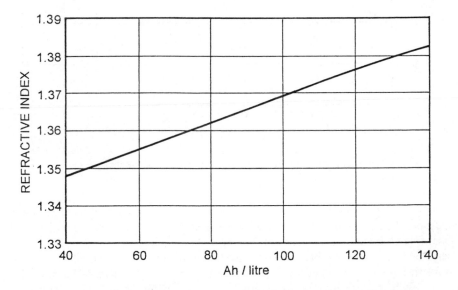

Fig. 5.14b Refractive index as a function of theoretical electrolyte capacity

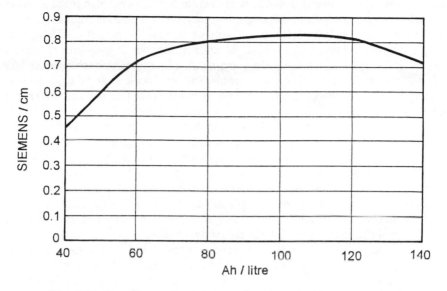

Fig. 5.14c Specific conductivity as a function of acid concentration

5.7.2 Electrical Methods
Electrical methods of measuring state of charge have some distinct advantages over physical types, namely:

a) Measurements may be made on a battery of cells, and changes in any individual cell can usually be detected.
b) No insertions of sensing devices need be made in the battery structure, and retrofit applications can be made.
c) The measurements are not prone to structural interference from bubbles or cell debris, as in the case of many physical sensors.

An obvious method of measuring charge state is to integrate current flowing into and out of the battery. Developments in electronics have made this easier and cheaper in recent years and it works fairly well for continuous changes between charge and discharge if the battery is never too near full-charge or overcharge. In the latter region, especially at low temperatures, charging inefficiencies make simple integration inaccurate. A further problem in the automotive application is obtaining a current sensor that retains its accuracy over the large range of currents involved (hundreds of amps to a few milliamps).

One useful and accurate measure of state of charge is the open-circuit voltage. Unfortunately this takes a long time to settle down after charge or discharge, and 24 hours are normally needed for a really accurate reading. If the battery has been overcharged, an even longer period is necessary, as anomalously high readings occur due to the presence of highly oxidising substances (peroxides and persulphates) in the electrolyte. Nevertheless, open-circuit voltage is only marginally affected by temperature and is often used to check the state of batteries in long-term storage. A variety of charge sensors have been marketed which claim to compensate for polarisation effects and calculate the equivalent of open-circuit voltage, thereby relating to state of charge. Their accuracy is often limited to specific duties and applications.

In recent years, many efforts have been made to measure the battery impedance at a variety of frequencies, and to relate the impedance to state of charge (Hughes et al, 5.4). Some limited success has been claimed in the case of stationary valve-regulated batteries. Impedance measurements are often used in specialised circumstances to measure the parameters of electrochemical reactions (Bard and Faulkner, 5.5). However, the impedances of larger cells (tens of Ah) are relatively small compared to interfering effects, and attempts to follow them in a dynamic automotive situation have met with limited success to date.

The development of an accurate battery model would allow measured current and voltages to be compared with stored model values and related to state of charge. Perhaps the ideal state of charge measurement awaits the development of a suitable mathematical or electrical analogue model which can be addressed in real time by a microprocessor.

5.8 Battery Modelling

As noted above, an accurate mathematical model is necessary for state of charge, or state of health, measurement by electrical means. (State of health refers to measurement of the decay of battery capacity with use.) Such a model is also an important feature of mathematical simulation programmes used in the design of automotive electrical systems (Holt, 5.6). Various attempts are being made to formulate a detailed mathematical model of individual cell or complete battery behaviour and to relate voltage/current responses to state of charge, temperature and battery life-history. Some of these were initiated by the needs of electric vehicle batteries in a different duty (Jayne ane Morgan, 5.7) but others have related directly to the automotive application (Holt, 5.8).

Mathematical modelling is frequently based on equations developed by Shepherd (5.9) which relate voltage to current and state of charge.

$$E = Er - Li. - K(^{Q}/_{(Q-it)})i. + A\exp(^{Bit}/_{Q})$$

where

A and B are constants characteristic of a particular battery cell
E = Voltage at time t at a steady current i
Er = Open circuit voltage
Li = Resistive voltage drop (iR)

K = Polarisation resistance per unit surface

Q = Capacity per unit area

(The constants may be determined by discharging in well-defined conditions at a steady current.)

Shepherd's equations tend to work most satisfactorily for steady discharges, whereas the automotive duty has varying discharge rates and varies continually between charge and discharge. Such a rapidly varying duty shows up the time-dependent aspects of polarisation behaviour and the capacitive behaviour of the battery cell.

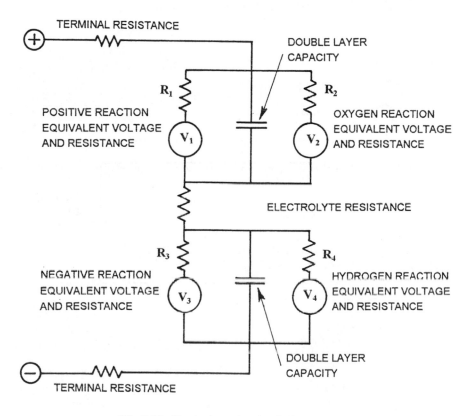

Fig.5.15a Equivalent circuit of a battery cell

A second approach is to model the cell or battery in terms of electrical circuit components as shown in Figures 5.15a and b. Such models can then be integrated into complete electrical systems and analysed by standard circuit analysis software. A

140

suitable equivalent circuit of a battery cell, based on individual electrode processes (Schoner, 5.10) is shown in Figure 5.15a. It is, however, difficult to obtain the values of the individual elements in Figure 5.15a, and the simplified version of 5.15b allows easier relation of the model elements to measurable parameters.

The ideal voltage sources, V_1, V_3, together with resistors R_1, R_3, represent the positive and negative electrode reactions; and V_2, V_4, with R_2, R_4, represent the oxygen and hydrogen evolution respectively. Since V_1 differs from V_2 and V_3 differs from V_4, there will always be self-discharge currents flowing in the individual electrodes. Both voltage and resistance values can be related to state of charge and temperature. Again the actual parameters are best determined by a series of characterisation discharges in fixed conditions.

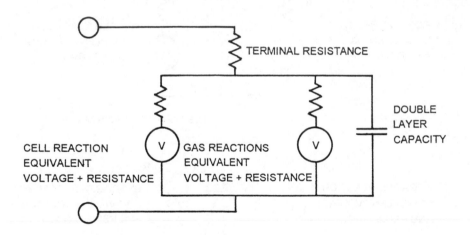

Fig.5.15b Simplified equivalent circuit of a battery cell

Battery life-limiting effects have received less attention where modelling is concerned. However, one useful relationship due to Seiger (5.11) estimates the effect of depth of discharge on cycle life.

$$Ld = Lo \exp k(1-d)$$

where L is the cycle life at depth of discharge d (expressed as a fraction, <1) and Lo is the cycle life at 100% rated capacity (i.e. with d=1). The constant k is a function of battery type and construction.

No totally satisfactory models have yet been published, but with the advent of automated test techniques and the use of circuit analytical software capable of resolving more complex models, progress in this field is expected to be rapid. An accurate model will then allow cheap microprocessor control of a variety of batteries, and continuous readout of the charge remaining, and running times, at various discharge rates.

REFERENCES

5.1 'A Fibre Optic Density Sensor for Monitoring State of Charge of Lead Acid Battery', G P Hancke, IEEE Transaction on Instrumentation and Measurement, 39 (1990), 247-50.

5.2 'In Situ Conductance Measurements of Lead Negative Plates', M Calabek, J Sandera, K Micka, Journal of Power Sources 10 (1983), 271-78.

5.3 'A Sensor for Specific Gravity of Lead Acid Electrolyte', M Tsubota in 'New Materials and Processes Vol.3 (1985)', pub. JEC Press Inc. Cleveland Ohio, 1985.

5.4 'The Estimation of the Residual Capacity of Sealed Lead-Acid Cells Using Impedance Technology', M Hughes, R Barton, S A G R Karunathilaka, N A Hampson, Journal of Applied Electrochemistry 16 (1986), 555-64.

5.5 'Electrochemical Techniques; Fundamentals and Applications' by A J Bard and L R Faulkner, Wiley Interscience, New York, 1979.

5.6 'Design Tools for Electric Power Systems' by M J Holt, AUTOTECH '89, Birmingham, UK, 1989 (Inst. Mech Eng. Paper C399/036).

5.7 'Novel Voltage Models of Lead Acid Battery Systems for Electric Vehicles', M G Jayne, C Morgan', Proc.EVS9, Toronto, November 1988 (paper EVS88-057).

5.8 'Predicting Load Balance and Electrical System Performance', M J Holt, 7th Intnl. Conference on Automotive Electronics, London, 1989 (I Mech Eng C391/10).

5.9 C M Shepherd, J Electrochemical Society, 112 (1965), 252-58 and 657-64.

5.10 'Electrical Behaviour of Lead/Acid Batteries during Charge, Overcharge and Open Circuit', H P Schoner, Paper EVS 88-063, Proc. EVS 9 Electric Vehicle Conference, Toronto, 1988.

5.11 'Effect of Depth of Discharge on Cycle Life of Near Term Batteries', H N Seiger, Proc.16th IECEC Meeting, Atlanta, Georgia, 1981.

CHAPTER 6

Nickel-Alkaline Batteries

Though the automotive field is dominated by the lead-acid battery, some penetration of the automotive market has been achieved by alkaline systems, namely the nickel-cadmium battery in its pocket-plate form. The advantages of nickel-alkaline batteries are usually summarised as follows:

- High power for a given capacity. The power capability does not diminish at lower states of charge, as is the case with lead-acid batteries.

- Corrosion of the conducting and support materials is minimal in the alkaline electrolyte and this contributes to a long life-expectancy. It also allows a greater choice of materials for the cell terminals, electrode supports and containers than does concentrated acid electrolyte.

- Only the water in the electrolyte is normally involved in the charge and discharge reactions (a mere 0.33 ml is required per Ah). This allows electrolyte volumes to be minimised without having significant concentration changes during operation. In addition the conductivity is not seriously reduced at low temperature.

- Due to the chemistry involved, it is easier to produce low-maintenance and sealed cells than in systems with acid electrolytes.

Nickel-cadmium batteries have therefore been adopted by the owners of vehicles with large electrical loads, particularly fleet-owners of public-service vehicles (e.g. buses) where the overall cost of ownership is more important than the initial capital cost and where, with proper maintenance, the longer life benefits can be realised. In recent years, improved lead-acid batteries have tended to displace alkaline batteries from even these applications and the market for automotive alkaline batteries has shrunk. There is, however, some possibility that new alkaline battery constructions and low-maintenance designs may bring new niche-applications in the automotive field and it is felt appropriate to consider the characteristics of these batteries in this volume.

This chapter will concentrate on the nickel-cadmium system, and particularly on the forms applicable to road vehicles. Readers interested in more detail will find the key text by Falk and Salkind (6.1) a useful source of specialised information on a variety of alkaline battery systems. Descriptions of smaller sealed alkaline batteries are contained in the text by Berndt (6.2).

6.1 Historical and Applications Background

Alkaline batteries with nickel hydroxide positives were developed independently and almost simultaneously by Thomas Edison in the USA and Waldemar Jungner in Sweden (Schallenberg, 6.3). Edison was searching for improved electric vehicle traction batteries. His US patent is dated July 16th 1901 (but the German version, valid from the application, is dated February 6th 1901) (Edison, 6.4). Jungner was interested in compact and reliable batteries for some early fire-alarm systems and was also attracted by the idea of large-scale electric power storage. His key Swedish patent was valid from January 22nd 1901(Jungner, 6.5) As might be expected, there was patent litigation between Edison and Jungner which contributed to the collapse of the original Jungner Accumulator Company. However, a new company was formed (Nya Ackumulator AB Jungner) which concentrated on the use of cadmium as the negative active material.

Edison produced nickel-iron cells for traction, although the first version of 1904 was withdrawn and replaced by an improved one in 1908. Nickel-iron cells powered several types of commercial electric vehicles in the early years of this century but are now no longer available, being replaced by the more powerful and efficient nickel-cadmium types.

Nickel-cadmium cells are now familiar items in a wide variety of applications. Small sealed cells (0.1 to 4 Ah) are used in many portable domestic and industrial appliances. Large vented cells (10 to 1000 Ah) are used in a number of stand-by power systems such as uninterruptible power supplies for computers. Special high-power forms (15 to 40 Ah) are used for the starting of aircraft engines, and low-maintenance forms (25-200 Ah) are commonly used for starting large diesel engines in both stationary and transport applications.

6.2 Chemical Principles

The positive active material of the alkaline systems considered here is hydrated nickel hydroxide. Its stoichiometry is complex, involving trivalent and quadrivalent nickel ions and a generalised formula of $xNiO_2.yNiOOH.zH_2O$ has been proposed (Bode, Dehmelt & Witte, 6.6). It is most frequently simplified to NiOOH. Discharge involves migration of protons into the hydroxide lattice to reach a structure equivalent to $Ni(OH)_2$. The discharge reaction is usually written:

$$NiOOH + H_2O + e^- \rightarrow Ni(OH)_2 + OH^-$$

(6.1)

In general, it is found that somewhat more than one electron is exchanged per nickel atom and the stoichiometry depends on the number of protons that can be incorporated into, and removed from, the particular nickel hydroxide lattice (de Wils and White, 6.7). The charged form and discharged forms of the active material can exist in several similar crystal structures but with different free energies and electrode potentials.

The charged material is unstable at high temperatures, giving an increased rate of self-discharge as it reverts to $Ni(OH)_2$ with the evolution of oxygen, and the discharged material is also difficult to recharge at higher temperatures.

$$4NiOOH + 2H_2O \rightarrow 4Ni(OH)_2 + O_2 \tag{6.2}$$

Tuomi (6.8) noted various forms of nickel hydroxide, i.e. the discharged material, depending upon the preparative procedure. Similarly, different types of the active material NiOOH exist depending on preparation route and on the existence of small quantities of stabilising elements which may be inserted in the crystal lattice.

Bode and co-workers (6.6) related the compounds formed at the positive electrode by the diagram shown in Figure 6.1. The changes between the various forms which occur on cycling result in a swelling of the active material which must be contained by the electrode and cell designs.

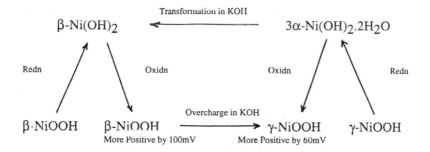

Fig.6.1. NiOOH: different forms and redox behaviour

In the highest oxidation state, some potassium (from the potassium hydroxide electrolyte) is incorporated in the lattice. Lithium ions may also be included in a similar manner and impart better charge-acceptance to the active material (Kelson et al, 6.9). For this reason, it is common to add several percent of lithium hydroxide to the nominally potassium hydroxide solution used as electrolyte. Cobalt additions to the active material also have a beneficial effect in slowing down the oxygen evolution on the material, thereby improving charge acceptance (Zimmerman and Efka, 6.10).

In some forms of cell, carbon in the form of graphite is added to the active material to provide better conductivity. (The nickel hydroxide material is a semiconductor whose conductivity decreases on discharge.)

It is obvious that a variety of additions can affect the behaviour of the variable-stoichiometry material that is formed on charging the positive electrode. New investigative techniques (e.g. neutron diffraction) are continuing to give more information about the precise structures involved, but the positive material still tends to be specified more often by its preparation route than by any exact chemical structure.

The negative electrode reactions are much more straightforward:

$$Fe + 2OH^- - 2e^- \rightarrow Fe(OH)_2 \text{ (nickel-iron)} \tag{6.3}$$

$$Cd + 2OH^- - 2e^- \rightarrow Cd(OH)_2 \text{ (nickel-cadmium)} \tag{6.4}$$

There are a variety of additives which may be included in the negative electrodes, particularly in the case of iron where copper and sulphur may be added to improve capacity. In earlier forms of cadmium electrode, iron was added to prevent agglomeration of the high surface-area form of cadmium and thereby maintain capacity. The end of discharge is determined by the coverage of the cadmium surface with a layer of cadmium hydroxide of critical thickness. Capacity is therefore dependent on maintaining the highest possible surface area of cadmium. Similar effects are achieved by some organic additives such as polyvinyl alchohol (Munshi and Tseung, 6.11). Nickel was also added with the idea of providing extra conductivity and limiting the maximum on-charge voltage. The nickel provides sites for hydrogen evolution, thereby sinking some of the charge current and preventing high negative potentials from being reached.

As may be expected from the previous discussion, the thermodynamic data for nickel positive material is variable, but an accepted value of -561 kJ/mole for the free energy of formation for β-NiOOH gives the free energy change for the cell reaction

$$2NiOOH + Cd + 2H_2O \rightarrow 2Ni(OH)_2 + Cd(OH)_2 \tag{6.5}$$

as -255 kJ/mole and the resulting equilibrium voltage as 1.32 V. The individual electrode potentials (calculated on the same basis) are:

$$NiOOH + H_2O + e^- \rightarrow Ni(OH)_2 + OH^- \qquad +0.51 \text{ V S.H.E.}$$

$$Cd + 2OH^- - 2e^- \rightarrow Cd(OH)_2 \qquad -0.81 \text{ V S.H.E.}$$

(That is, with reference to the Standard Hydrogen Electrode)

These can be compared with the oxygen and hydrogen evolution potentials in alkaline solution:

$$4OH^- - 4e \rightarrow O_2 + 2H_2O \qquad\qquad +0.40 \text{ V S.H.E.}$$

$$2H_2O + 2e \rightarrow H_2 + 2OH^- \qquad\qquad -0.83 \text{ V S.H.E.}$$

The cadmium electrode is thus stable in the alkaline solution. Polarisation effects on charging drive it more negative than the hydrogen potential. However, the hydrogen evolution reaction is fairly slow on the cadmium surface and a large kinetic effect prevents any appreciable hydrogen evolution taking place unless the electrode is shifted to potentials several hundreds of millivolts more negative, as actually happens at the end of charge. This implies that the cadmium charging is fairly efficient, most of the current being directed to the reduction of the cadmium hydroxide.

The iron electrode has an equilibrium potential of -0.56 V, which is well above the hydrogen potential. However, hydrogen evolution occurs at a significant rate on its surface, and once the electrode is forced below the hydrogen potential on charging, gas evolution readily occurs and current is wasted in the hydrogen evolution reaction. This was reflected in the generally poor electrical efficiency of nickel-iron cells and partly accounts for their demise in favour of nickel-cadmium. The rest of this chapter will concentrate on the nickel-cadmium battery.

The oxygen evolution reaction on the nickel hydroxide electrode is appreciably fast. As noted earlier, additives to the nickel hydroxide lattice can decrease the oxygen evolution rate and thereby keep the majority of the current in the charging reaction rather than being diverted to oxygen evolution. Even so, the charging efficiency decreases as the electrode reaches full charge and significant oxygen evolution takes place. This situation is exacerbated at temperatures much above 40°C, and charging efficiency drops markedly.

In contrast to the lead-acid cell, the maximum utilisation of the active materials in alkaline cells tends to be high, with the theoretical values indicated below (based on a one-electron change between NiOOH and $Ni(OH)_2$).

$$2NiOOH + Cd + 2H_2O \rightarrow 2Ni(OH)_2 + Cd(OH)_2 \qquad\qquad (6.6)$$

		(discharge)
NiOOH	3.42	g/Ah
Cd	2.10	
H_2O	0.67	
$Ni(OH)_2$	3.46	
$Cd(OH)_2$	2.73	

These values give a theoretical energy capacity of 215 Wh/kg (compared to 176 Wh/kg for lead-acid batteries).

The theoretical figure should be compared with the practical figures of 25-55 Wh/kg which are usually achieved. In practice, although the utilisation of the positive material is very high, only about 70% of the cadmium theoretical capacity is achieved. In addition, the weights of the conductive supports and the other inert parts of the electrode structures are also high and detract from the energy per unit-weight actually achieved.

Oxygen reacts rapidly with the metallic cadmium of the negative electrode and, as with the lead-acid system, it is possible to construct sealed cells using an oxygen recombination cycle (in fact the gas-recombination cycle was first used with nickel-cadmium since the oxygen-recombination is faster on the wet cadmium surface). The sealed cells are constructed with an excess of negative active material so that overcharge always results in oxygen evolution unless very heavy and prolonged overcharge takes place. The oxygen produced must be allowed access to exposed cadmium and this may be achieved by absorbing the electrolyte on the electrodes and separators so that the cell contains no free liquid. With the correct amount of electrolyte and with suitable absorbent separator materials, the oxygen may then freely diffuse to the cadmium electrode surface. An alternative strategy uses cadmium electrodes with an internal wetproof layer of porous PTFE which allows gas access to the centre of the cadmium active mass (Crompton, 6.12). The reactions involved on overcharge are then as follows.

At the Positive $\qquad$ $4OH^- - 4e \rightarrow O_2 + 2H_2O$ $\qquad$ (6.7)

At the Negative $\qquad$ $O_2 + 2H_2O + 2Cd \rightarrow 2Cd(OH)_2$ $\qquad$ (6.8)

$$2Cd(OH)_2 + 4e \rightarrow 2Cd + 4OH^- \qquad (6.9)$$

A schematic of the cycle of reactions taking place within the cell is shown in Figure 6.2. The magnitude of the overcharge current which can be accommodated by this means depends on the gas-diffusion to the cadmium and the dissipation of the heat released by the recombination reactions. Oxygen recombination at the negative

POSITIVE PLATE $\qquad$ **NEGATIVE PLATE**

CHARGE $\qquad$ *CHARGE*

$Ni(OH)_2 + OH^- - e \rightarrow NiOOH + H_2O$ $\qquad$ $Cd(OH)_2 + 2e \rightarrow Cd + 2OH^-$

OVERCHARGE $\qquad$ *RECOMBINATION*

Oxygen Diffusion to Negative→

$4OH^- - 4e \rightarrow 2H_2O + O_2$ $\qquad$ $2Cd + O_2 + 2H_2O \rightarrow 2Cd(OH)_2$
$Cd(OH)_2 + 2e \rightarrow Cd + 2OH^-$
(Also direct oxygen reduction
$O_2 + 2H_2O + 4e \rightarrow 4OH^-$)

Fig.6.2. Schematic of charging reactions in a sealed cell

electrode surface causes a positive shift in the potential of that electrode which is manifested in a slight decrease in cell voltage. Since the cell voltage rises during the normal charging process, this slight decrease may be used to signal that the cell is fully charged and gas evolution is starting. Modern battery management circuits for sealed cells often take advantage of this phenomenon in determining the point of full charge.

6.3 Nickel-Cadmium Battery Construction and Manufacture

As noted earlier, the wide range of materials available for supports in alkaline electrolytes has produced a variety of cell constructions. Cells are generally classified by the type of support used for the electrodes' active materials as follows:

a) Pocket-Plate (or Tubular) Cells
The earliest type of construction (developed by Jungner) producing rugged, long-life electrodes which contain the active materials in perforated steel pockets or tubes giving good overall performance but with some weight and cost penalties.

b) Sintered-Plate Cells
A construction with electrodes based on high-porosity sintered nickel plaques impregnated with the active materials. Capable of very high power outputs and amenable to sealed-cell constructions. Materials and manufacturing costs are high.

c) Plastic-Bonded Cells
The latest type of construction using composite plaques of active material and a conductive support (usually graphite) together with a plastic binder. The plaques are roll-bonded onto sheets of nickel or other corrosion-free conducting material. Alternatively, a paste of active material and binder may be applied to a sheet of inert conducting material and dried and cured in situ. Low-weight constructions with good performance are produced but the life may be limited by gradual changes in the conductive filler or the binder. The life of positive electrodes which use graphite as a conductive support, in particular, tend to be shorter than with the other types of nickel-cadmium cell.

Recently, the French company SAFT has marketed cells using sintered positive electrode constructions and plastic-bonded negative electrodes. These give high performance at reduced cost compared to normal sintered cell constructions. They also have advantages in improved life of the cadmium electrodes for reasons which are described later.

There are also several variations on the sintered electrode theme. The porous plaques made from sintered nickel powder are replaced in some designs by nickel-plated steel fibres sintered together to form a porous mass (e.g. those manufactured by the German company Hoppecke).

Batteries for automotive purposes traditionally use the pocket-plate construction and this will be described in more detail than the others. Sealed cells with sintered electrodes are now used in a wide variety of smaller forms in high value portable appliances where the high power and convenience of sealed construction are important and the battery cost is not too significant. Plastic-bonded electrode technology has made limited headway so far in alkaline cells except for the SAFT bonded negatives noted above. The bonded positive awaits further material developments for more stable conductive supports than the graphite powders currently available.

6.3.1 Pocket Plate Cell Construction
Figure 6.3. shows the construction of a pocket-plate cell and Figure 6.4. shows the detailed construction of one of the electrodes.

Strips of thin steel ribbon 5 to 10 mm wide are perforated with small holes (0.1 to 0.5 mm) which account for 12 to 18% of the total surface. The strip is then plated with nickel (2.5 to 4 μm thick) to prevent corrosion. Plating may be followed by annealing in a reducing atmosphere to ensure good plating-adherence in later treatment. Two strips of the perforated ribbon are then pressed onto a compacted strip of the active material and the strips are rolled to produce a seal at the edges, thereby enclosing the active material in a perforated rectangular-section pocket. The pockets are welded side-by-side to an electrode frame (also of nickel-plated steel) which has a lug or lugs to make contact with the appropriate cell terminal. The process is shown in Figure 6.5.

Pocket dimensions affect the resistive path and the total surface available, hence having a major effect on the discharge performance. High-rate cells use thin and

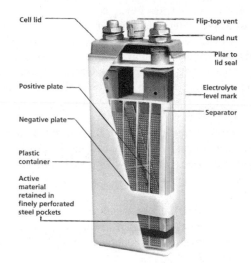

Fig.6.3. Pocket plate cell construction
(Alcad Ltd)

narrow pockets (and hence more plates and higher surface area for a given capacity). Low-rate cells can use wider and thicker pockets with a more economical construction.

A variation due to Edison used tubes made by spiralling the perforated strip to contain the positive material instead of the rectangular pockets, but such constructions are no longer used.

The active materials are pressed into rectangular tablets or strips before enclosure in the pockets as described above. The positive mix uses nickel hydroxide (produced by precipitation with sodium hydroxide from nickel sulphate solution). Cobalt hydroxide is co-preciptated in a proportion of 5% by weight, for the reasons discussed in Section 6.2. Several percent of graphite or nickel flake is then added to improve the electronic conductivity, together with a small amount of paraffin as binder in the pressing operation. As noted earlier, the exact preparation conditions for the nickel hydroxide have a profound effect on performance, and the material purity and crystallite size are carefully controlled.

The negative mix consists of precipitated cadmium hydroxide, sometimes with iron additions (to prevent growth of the cadmium particles and consequent loss of active surface on cycling) and again with a paraffin binder. Graphite may be added to assist the pressing operation. Cadmium oxide may also be used as a starting material.

The pocket plates are assembled into cell packs using a variety of separators. High-power cells may have plastic rods as spacers between the individual electrodes. Otherwise, perforated corrugated plastic, or woven plastic thread, may be used. No particularly critical properties are required of the separator in this type of cell construction.

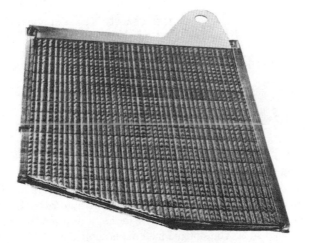

Fig.6.4. Pocket plate cut-away
(Alcad Ltd)

152

The plate connections at the top of the cell may be made by welding the plate frame lugs together, or more often by bolting the lugs together with nickel-plated steel washers or spacers (Figure 6.3). High-rate cells may have two or more cell terminals for each set of plates. The terminals are made of heavily nickel-plated copper or steel, depending on the application.

Cell boxes can be made from a variety of rigid polymers but polypropylene is favoured; thermal welding can be used for the sealing of the cell tops. Occasionally cells are supplied in coated steel cases for added mechanical protection. Pocket plate cells are normally provided with individual cell vent-caps incorporating positive-pressure relief-valves.

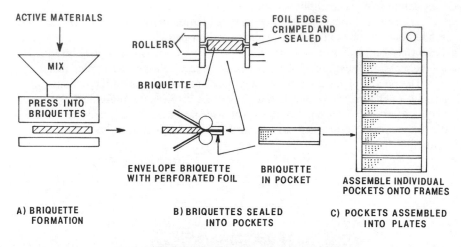

Fig.6.5. Production of pocket plate electrodes

Batteries of pocket plate cells have in the past been enclosed in rigid wooden or painted steel boxes. More recently, a variety of plastic moulded modules have been used (Figure 6.6).

6.3.2 Sintered Cell Construction
Much of the skill (and expense) in sintered cell construction rests in the preparation of the electrode support plaques. These consist of layers of sintered nickel up to about 1 mm thick on either side of a sheet of nickel wire mesh or, more usually, perforated nickel or nickel-plated steel. The sintered material is applied as a slurry (in water with an appropriate binder such as carboxymethyl cellulose) to a perforated sheet of nickel or nickel-plated steel. This layer is then dried in air and sintered (at 800-1000°C) in a reducing atmosphere (e.g. the nitrogen/hydrogen mixture produced from cracked

ammonia). A carefully selected combination of nickel particle sizes and sintering conditions must be used to produce sintered layers with very high porosity and internal surface area yet with good mechanical integrity and good electrical conductivity. This particular combination of properties involves a careful balance of conflicting requirements. The slurry application and sintering may be configured as a continuous process as shown in Figure 6.7.

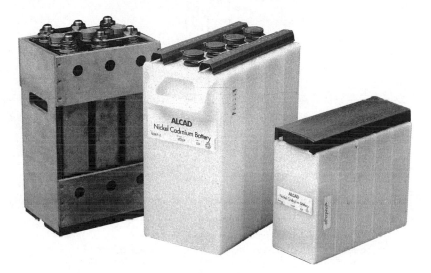

Fig.6.6. Plastic moulded modules
(Alcad Ltd)

Typical sintered layers have porosities of 80-90% and pores of 5-15 μm diameter. Internal surface areas, as measured by the BET method, amount to 0.25-0.5 m^2/g, being highly dependent on sintering time and temperature.

The strip with sintered layers is then coiled into open spirals and the sintered layers are impregnated with active materials by one of a variety of methods, the main two being described as follows.

a) *Chemical Impregnation*. Active materials are formed within the pores of the sintered layer by vacuum impregnation with, alternately, a nickel salt solution and sodium hydroxide solution. Nickel hydroxide is thus deposited inside the pores of the positive plates. A similar treatment with cadmium salt solution and potassium hydroxide solution is used to precipitate cadmium hydroxide in the pores of the negative plates. Nickel nitrate and cadmium nitrate are the salts normally used. A few percent of cobalt nitrate is added to the nickel nitrate solutions for the positives in order to improve performance, as noted earlier.

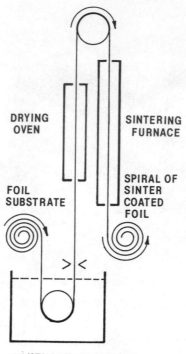

Fig.6.7. Schematic of production of foil coated with sintered nickel layer

b) *Electrochemical Impregnation*. The porous nickel layers are vacuum-impregnated with nickel or cadmium salt solutions and then made cathodes in a potassium hydroxide solution. In addition to the hydroxide precipitation by the electrolyte solution, hydrogen evolution taking place on the plates creates a stirring effect which improves hydroxide penetration into the pores. Hydrogen evolution on the electrodes also produces more hydroxyl ions within the pores.

$$2H_2O + 2e \rightarrow 2OH^- + H_2$$

These effects produce more effective precipitation of the hydroxides within the pores of the sintered material.

In a more recent development in the USA (the Kandler Process), electrolysis takes place in nitrate solutions of the appropriate salts with pH controlled in the slightly acid or neutral region. The hydroxyl ions produced on electrolysis then precipitate the hydroxides within the pores of the solution (Figure 6.8).

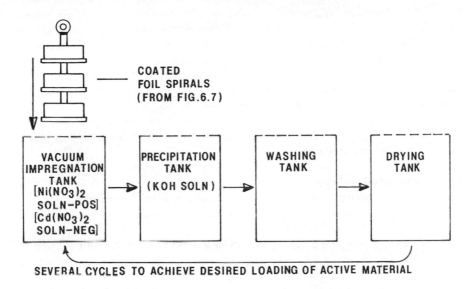

Fig.6.8. Chemical impregnation of nickel sinter

In both cases, the treatment may be repeated several times to achieve the desired loading of active materials, and various proprietary improvements have been made to reduce the number of cycles and time required. There also follow several washing stages to remove any unreacted soluble nickel (or cadmium) salts. Nitrate remaining in the plates can result in self-discharge of the completed cells since it is reduced on the negatives to nitrite, which diffuses to the positive and is re-oxidised to nitrate (commonly referred to as the 'nitrate shuttle' mechanism).

The active materials formed in the plates as prepared are both in the discharged forms and require one or more cycles of charge and discharge to arrive at the final product.

Sintered cells can have similar plate interconnection and terminals to pocket plate cells. However, the higher discharge rates achieved with sintered cells usually make it advantageous to use welded electrode connections and nickel-plated copper terminals.

The separators used in sintered cells are usually of more advanced types. Porous polypropylene (Celgard™ by British Celanese) and a variety of materials based on radiation-grafted polymers are often used. The latter rely on ion-exchange processes for ion transfer. The improved separators allow very small interelectrode spacing giving reduced cell resistance. At the same time they prevent (or rather retard) dendrite growths from the cadmium electrodes from causing short circuits. Though polypropylene separators are very stable, they are naturally hydrophobic and need either chemical surface treatment or radiation grafting of hydrophilic groups to ensure that they are properly wetted out.

In tightly packed cell assemblies, absorbent layers of unwoven polyamide fibre may also be used as electrolyte reservoirs. This material is not completely stable in strong alkali, especially at high temperatures, but serves well when no particular mechanical strength is required.

Sintered plates are more prone to positive active material changes than pocket plates, and the tightly packed cell constructions often swell visibly with cycling. These swelling effects have to be accommodated in cell and battery construction. Batteries are thus mounted in rigid cases of metal or plastic (polypropylene) reinforced with wood or metal. However, specially designed reinforced wall plastic cases are now being introduced.

6.3.3 Plastic-Bonded Plates

Plastic-bonded cadmium plates are made by coating a nickel (or nickel-plated steel) foil with a layer of paste consisting of cadmium oxide powder and a suitable inert polymer binder (SAFT). An alternative process (Jindra et al, 6.13) consists of calendering a mix of cadmium oxide and binder to form a dough and rolling layers of the resulting dough onto a conductive supporting foil. When the plates are charged, the cadmium oxide is reduced to a high surface area layer of porous cadmium metal. The layers of active material may be up to several millimetres thick but the utilisation of active material decreases with thicker materials at high discharge rates.

Several attempts have been made to apply similar processes to the nickel positive plates but have been hampered by the volume changes in the active material as it changes form (Section 6.2) and by oxidation of graphite powder used as a conductive support. At the time of writing no plastic-bonded positive plates are used in commercially available cells.

6.4 Operating Characteristics Performance of Pocket Plate Nickel-Cadmium Cells

Figure 6.9 shows the changes in electrode potentials and in cell potential for a typical pocket plate cell (with excess electrolyte and vented so that oxygen recombination does not occur). Cells are normally constructed with some excess cadmium so that capacity is limited by the nickel hydroxide. The utilisation of the nickel hydroxide may be 100% (or even slightly more, as explained in Section 6.2) based on a theoretical one electron change per nickel atom.

A particular feature of nickel-cadmium cells is the low reduction in effective capacity at high discharge rates, as shown in Figure 6.10.

The effect of temperature is shown in Figure 6.11. Ordinary pocket plate cells can be used down to -25°C, when the electrolyte begins to freeze. More concentrated electrolyte allows operation down to -50°C. Operation above 45°C may produce permanent damage in the positive material.

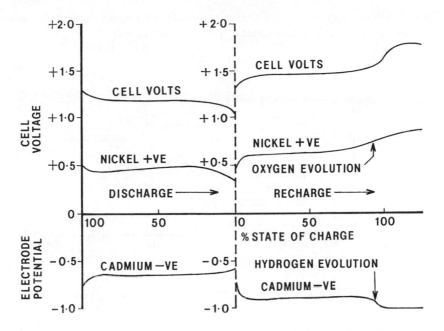

Fig.6.9. Cell voltage and electrode potentials for nickel-cadmium pocket cell (0.2C rates)

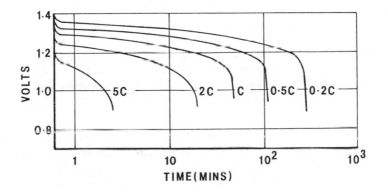

Fig.6.10. Cell voltage/time at various discharge rates

6.5 Operating Characteristics and Performance of Sintered Plate Nickel-Cadmium Cells

Sintered plate cells are characterised by higher voltages and better capacity maintenance at high discharge rates. Rates of 10C or even 20C may be provided for

158

limited periods by cells with appropriate terminals and internal connections. Figure 6.12 shows the discharge performance of a typical sintered cell.

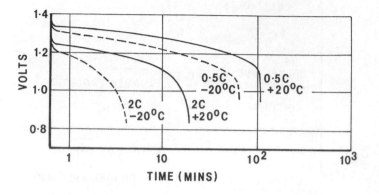

Fig.6.11. Effect of temperature on discharge voltage (high-rate cell)

6.6 Charging of Nickel-Cadmium Cells

The preferred recharging of nickel-cadmium cells is with constant current at rates between C and C/10 to give a fixed fraction of overcharge in excess of the charge removed. The fraction is usually between 20 and 40% and is required due to the inefficiency of the positive recharging process. Since a stoichiometric excess of negative material is provided, this is not completely charged until the positive recharge is complete.

As full charge is approached, oxygen is first evolved freely due to the positive inefficiency, and as the cadmium full charge is approached hydrogen is finally evolved in fair amounts in addition to the oxygen. However, overcharge may have to proceed for some time before the stoichiometric mixture for water splitting (i.e. $2H_2 + O_2$) is reached.

As with lead-acid cells, nickel-cadmium cells can also be float-charged at constant voltage. The voltage chosen has to be a compromise between high values required for rapid and complete recharge and lower values preferred for low gas evolution and water loss. Pocket plate high-rate cells, of the type used for automotive duties, have recommended on charge voltages of 1.55 to 1.6 V/cell in an automotive duty. Table 6.1 shows the effect of the charging voltage limit on the cell behaviour.

6.7 Life-Limiting Effects

As noted earlier, nickel-alkaline batteries are free of the corrosion effects that can frequently limit the life of lead-acid electrodes and cells. In the absence of construction faults and assuming that electrolyte water loss is made good, the effects which control the life of cells are:

1. Cadmium dendrite formation
2. Cadmium capacity loss
3. Positive swelling effects
4. Positive capacity loss
5. Electrolyte carbonation

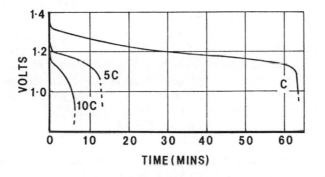

Fig.6.12. Discharge performance of sintered cell

Table 6.1. Effect of Charging Voltage Limit on Cell Behaviour

Voltage Limit (V/Cell)	Comments
1.40	Minimum float voltage for normal operation. Very low water loss but requires regular boost charges.
1.45	Recommended float voltage for normal operation. 80% of capacity returned in 5hrs after full discharge. Full charge requires >1 week. Minimal water loss.
1.53	Some water loss in continuous float operation but full capacity restored in <1 week.
1.58	High water loss in continuous operation, but full capacity restored in 24 hrs.
1.68	High water loss but will recharge in 8hrs if a current >0.2C is available.

Cadmium is not entirely insoluble in the alkaline electrolyte, particularly if carbonate is present, and the presence of soluble cadmium compounds results in the gradual growth of spiky cadmium formations or dendrites from the negative electrodes and across the separators until a short circuit is formed. The initial growths may only form a high resistance path, in which case an increased rate of self-discharge is noticed.

Conditions are worsened in a series chain of cells, since the failing cell may never become properly charged before the voltage limit of the full chain is reached. On cycling, the cell continues to deteriorate till its effective capacity is seriously reduced. Such dendrite growths are encouraged by prolonged overcharge, high temperature operation or poor electrolyte maintenance. The problem is more serious with sintered or plastic bonded cells which have thinner separators (dendrites will grow through many separators). It is sometimes claimed that a temporary cure can be effected by applying a very high charge current for a matter of seconds, the object being to burn away limited dendrite growth. However, the high currents required can cause damage to other parts of the cell.

Cadmium capacity loss occurs due a gradual growth of the cadmium crystallite sizes in the active material and the resulting loss in active surface area. Polarisation effects are thus higher for a given discharge current, and the end of discharge occurs when the surface is covered by a limiting-thickness layer of cadmium hydroxide.

Various additives (e.g. iron as already noted, but now more frequently organic materials such as polyvinyl alcohol) may be added to be absorbed onto the crystallite surfaces and retard growth effects (Barton et al, 6.14).

A further effect when nickel is present in intimate contact with the active material is the formation of cadmium-nickel alloys, e.g. Ni_2Cd_5 or Ni_5Cd_{21}. These are usually destroyed on complete discharge but may remain on partial discharge and result in lower cell discharge voltages.

As noted in Section 6.1, the positive active material swells significantly during use, and the pressures brought about within the cell can sometimes rupture separators and cause short circuits. Occasionally even the cell cases themselves can be broken.

As noted earlier also, the positive material is thermally unstable and higher temperature operation may cause slow degradation and capacity loss.

The electrolyte, being based on potassium hydroxide solution, readily absorbs carbon dioxide from the atmosphere to form potassium carbonate. The carbonate solutions have markedly lower conductivity and will eventually degrade high-rate performance. In addition, the presence of carbonate increases cadmium solubility and increases the chance of cadmium dendrite formation. Well designed cells are normally provided with airtight caps allowing only positive pressure venting and preventing any appreciable air ingress.

Which of these particular effects causes the demise of the cell depends very much on its particular duty and treatment. Well designed cells now incorporate the means to slow down all of the above effects by various proprietary chemical or mechanical methods. Cycle lives in excess of 500 complete discharge cycles can be expected for heavy duty industrial cells, and in lighter duties and shallower cycles, cycle lives of several thousands are achieved. When little cycling occurs and the cells are well maintained (as in some railway applications) nickel-cadmium batteries have been found to function satisfactorily over periods of twenty years.

6.8 Memory Effects

With sintered cells and certain of the other types where nickel supports or additions are added to the negative material, so-called 'memory effects' are observed in cells that are continually shallow-cycled or continually float-charged. The effects manifest themselves as marked drops in discharge voltage at the depth of discharge most frequently used, as illustrated in Figure 6.13 (or, in the case of continuous float-charging, a drop in voltage before the cell reaches anywhere near its rated capacity.

The memory effects observed are virtually eliminated by the complete discharge of the cell (at reduced voltage if necessary) and a full re-charge when most of the negative material is reformed in its high surface state. They are attributed to the effects explained above, namely cadmium crystallite growth in the part of the cadmium that is rarely discharged and gradual intermetallic compound formation on continuous charging (the intermetallic compounds discharge at more positive potentials than cadmium).

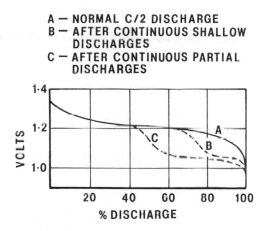

Fig.6.13. Memory effect

6.9 Environmental Considerations

Cadmium is a toxic heavy metal and most countries restrict its entry into the environment. Cadmium metal or metal oxide fumes can be readily absorbed in the lungs and cadmium salts may be ingested orally.

The cadmium contained in batteries is in a form which can never be released into the environment in normal battery operation. The metal and metal oxide or hydroxide is embedded in the battery structure and is in a stable solid form. Batteries

162

themselves are therefore relatively safe containers for this type of material.

Hazards may occur when batteries are crushed and incinerated with other refuse in a haphazard manner. Expired nickel-cadmium batteries of any size should be disposed of by returning to the suppliers or by properly qualified reclamation agents. However, the cost of the raw materials makes the recycling of nickel-cadmium batteries fairly attractive and many manufacturers now offer recycling facilities. Larger batteries, such as the types used in vehicles are particularly attractive from a recycling point of view and most of these are reclaimed when their useful life has finished. Nearly 100% of the cadmium content can be re-used. Some manufacturers are now also considering leasing larger batteries, thereby ensuring safe re-cycling.

REFERENCES

6.1 'Alkaline Storage Batteries' by S U Falk and A J Salkind, Wiley, New York,1969.

6.2 'Maintenance-Free Batteries' by D Berndt, Research Studies Press, Taunton, 1993.

6.3 'Bottled Energy' by R H Schallenburg, American Philosophical Society, Philadelphia, 1982.

6.4 T A Edison, German Patent 157,290(1901); US Patent 678,722(1901).

6.5 W Jungner, Swedish Patent 15,567(1901).

6.6 'Nickel Hydroxide Hydrate' by H Bode, K Dehmelt, J Witte, Proc. CITCE Meeting, Strasburg, 1965.

6.7 P de Wils, R White, J.Electrochemical Soc. 142(1995),1509.

6.8 D Tuomi, J.Electrochemical Soc.,112(1965),1.

6.9 'Behaviour of Sintered Plate Nickel Hydroxide Electrodes in Lithiated Electrolyte Solutions' by P Kelson, A D Sperrin, F L Tye, in ' Power Sources 4, Oriel Press, Newcastle 1973.

6.10 'Effect of Additives on Nickel Electrode Discharge Kinetics' by A H Zimmerman, P K Efka, in 'Power Sources 10', Ed. L J Pearce, Paul Press, London, 1985.

6.11 M Z A Munshi, A C C Tseung, J. Power Sources,18(1986),33.

6.12 'Small Batteries - Secondary Cells' by T R Crompton, (p.116), Macmillan, London, 1982.

6.13 'Plastic Bonded Electrodes produced by a Rolling Technique' by J Jindra, J Mrha, K Micka, Z Zabransky, B Braunstein, J Malk, V Koudelka, in 'Power Sources 6', Ed. D H Collins, Academic Press, London, 1977.

6.14 R Barton, S J Lawson, P J Mitchell, N A Hampson, J. Power Sources,18(1986),43.

CHAPTER 7

Future Developments

The lead-acid battery has performed well in the broad spectrum of automotive duties and, at the time of writing, seems unlikely to be displaced from its dominant position. A vast production investment, now complemented by recycling facilities, ensures a continuing grip on the battery markets for mass-produced vehicles. However, it is important to consider the influences for change, both in terms of vehicle requirements and in terms of battery technology. This chapter examines the way in which the market and technology developments might produce significant changes in the medium to long-term future.

7.1   Developments in Vehicle Systems

The significant trends in automotive electrical systems may be listed as follows:

1) Increase in total power demands due to larger numbers of electrically powered sub-systems with significant current requirements (e.g. windscreen and window heating, seat heating, electric pre-heating of catalytic exhaust converters).

2) Increase in load requirements when the engine is inoperative (e.g. alarm and security systems, communication and computing systems).

3) The need for greater system reliability due to electronic control and electric actuation of critical functions such as engine management, steering and brakes.

4) Changes in power requirements for key functions such as engine-starting (e.g. by the use of capacitors or flywheels for intermediate energy storage or by changes in engine technology).

5) Environmental forces bringing the use of battery-electric or hybrid vehicles. Electric traction systems demand the use of high voltages and the discussion of traction batteries is beyond the scope of this volume. However, such vehicles are still likely to use a low voltage sub-system for lighting and auxiliaries, similar to the existing automotive systems. The

alternator will be replaced by the traction source and a voltage-reducing DC/DC converter.

7.1.1 Higher Voltage Systems

Demands for increased power have grown steadily (see Chapter 1) and the first response of system designers was to consider the use of higher voltages (Jarrett and West, 7.1; Williams and Holt, 7.2). Systems based on a nominal 24 volts are already well established in large commercial and public service vehicles (i.e. buses and minibuses). These are generally successful, but 24 V filament lamps are more fragile than 12 V versions. Discussion of maximum safe working voltage has led to the proposal of nominal 48 V systems. These would reach 60 V or so during charging of the battery with full available output, and this value (i.e. 60 V) is regarded as the maximum voltage permitted without special protection. Some of the 48 V proponents would use an additional system at 12 V (fed from a DC/DC converter) to accommodate more robust 12 V filament bulbs, though this could be rendered unnecessary by gas-discharge lighting units and high-brightness light-emitting diodes now under development.

The adoption of higher voltages poses no formidable technical obstacles, since more cells, or more battery modules, may be placed in series to obtain the higher voltage required. However, larger numbers of cells have to be fitted in roughly the same space and, assuming that the overall battery energy and volume are not to be increased, a higher voltage battery will have a lower ratio of active to inactive materials. This is brought about by the larger number of cell partitions and interconnections required and the inevitable dead-space in each individual cell. Thus for a given total energy stored, high voltage batteries will have slightly lower energy densities and relatively higher costs than those working at lower voltages with fewer cells.

As the number of cells in series increases, other problems become significant. If one cell in a series chain is of slightly lower capacity than its neighbours, it will be more deeply cycled when the battery is operated between fixed voltage limits. This effect tends to reduce the life of a series chain of cells as compared to the life of individual cells. In the worst case an individual cell may be overdischarged to the point where its electrodes suffer polarity reversal (i.e. the electrodes begin to be formed up in the opposite polarity) before the chain as a whole reaches the discharge cut-off voltage. Naturally, the effect becomes more significant as the voltage, and hence the number of cells, increases. It is not particularly prominent in 12 V batteries of six cells in normal automotive duty. However, the effect would be exacerbated considerably with larger numbers of lower capacity cells. (The effect is pronounced in high voltage electric vehicle traction batteries, but special quality controls to reduce capacity dispersion and the use of battery management systems, which monitor small groups of cells in a series chain, help to diminish the problem.)

More recent reviews of the automotive situation have tended to discount higher voltages, not because of the battery problems, but rather due to other system disadvantages (e.g. with connectors and voltage transient suppression) which have offset the perceived advantages (Williams et al, 7.3). Whilst it is certain that the pressures now on electrical systems will bring significant changes, the exact voltages and the system architecture used will be the subject of considerable argument. The cost of changing from the present relatively simple 12 volt system and the need to establish new standards which are widely accepted imply a long period of debate and investigation before changes are implemented. In the meantime more intelligent use is being made of 12 V by imaginative use of new system architectures.

7.1.2 Multiple Battery Systems

As noted in Chapter 1, a vehicle may be immobilised by power system failure, and this is borne out by the statistics from UK motoring organisations showing that more than 50% of roadside failures are due to some aspect of electrical system behaviour. Modern vehicles, especially those with automatic transmission, have no guaranteed means of starting apart from the electric starter-motor. Moreover, as indicated earlier, the engine fuelling and ignition are also dependent on the battery. There is therefore a premium placed on being able to start the engine, which, when active, will provide electrical support from the alternator.

A variety of systems are being explored with the objective of an assured engine-start capability. These would ensure that the engine could always be started even after prolonged use of all electrical loads with the engine inoperative (e.g. after remaining stationary in a motorway traffic jam on a cold winter evening).

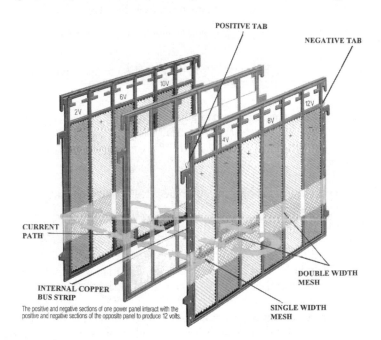

Fig.7.1. Dunlop Pulsar battery
(Gould National Batteries Inc.)

Initial developments have involved the use of dual batteries, i.e. the combination of two separate batteries in a single container. Examples have been marketed by GNB (as the Champion 'Switch' Battery) in the USA and by the New Zealand company Power Beat. Both of these involved a standard-sized battery together with a high power battery of smaller size in a common container. The actual arrangement is as shown in Figure 7.1. The Switch battery also used the special high power Pulsar™ construction developed by Pacific Dunlop Company (Howlett, 7.4), and a switch connecting both battery sections in parallel for emergency starting is part of the battery structure. This is accessed from the engine compartment and is meant as an emergency facility only for engine-starting when the main battery is too discharged to operate the starter unaided. The Power Beat version uses proprietary electronics to connect the sections in parallel automatically during the starting period.

Both of the above examples are designed for easy inclusion in existing vehicle systems. A number of automotive companies are now examining radical changes to vehicle electrical architecture which involve two or more batteries, one of which is dedicated to the starting function. The separation of starting and auxiliary load functions allows some improvements in battery design and perhaps some unexpected economies. If the starting battery is always fully charged, its capacity can be reduced and it does not need real cycling capability. The auxiliary load battery does not need the high power capability for engine-starting. A study by the French company CEAC (Douady et al, 7.5) notes the required characteristics as follows.

Starter Battery

- Ability to deliver full cranking power in the operating temperature range.
- Low weight.
- Low volume.
- Ability to operate in continuous overcharge conditions.
- Ability to operate in engine space near to starter.

Service Load Battery

- Capacity to support loads (especially quiescent loads) over long periods.
- Deep cycling capability.
- Ability to recover from storage at low states of charge.
- Low weight.
- Good charge acceptance in working temperature range.
- Ability to supply high current for a few seconds (e.g. to braking system pump).
- Preferably low-maintenance design allowing fitting in any part of the vehicle.

The electrical system architecture has to take into account that the batteries may be functioning at different temperatures and that long leads with corresponding voltage losses may exist between the alternator and the service-load battery. The starting battery must have priority of charging. A split system involving two battery current relays is suggested by CEAC (Douady, 7.5) as in Figure 7.2. The operation of the relays would be under electronic control by a system relating to the batteries' states of charge and the status of the charging system and load priorities.

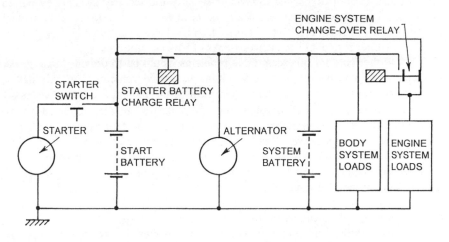

Fig.7.2. CEAC dual battery system diagram

At the moment, the most cost-effective candidate for both batteries is a lead-acid 12 V module. The starter battery may be a high-rate design of the types referred to later in this chapter and the service battery could be a VRLA (Valve- Regulated Lead-Acid) design with strengthened grid and plate structures to give deep cycling capability. The deep cycle capability of the nickel-alkaline battery would be useful in this function, but present costs would be prohibitive.

7.1.3 Quiescent Drain Control

In recent years the current drawn from the battery when the vehicle is out of use has shown a steady increase. The small milliamps leakage through alternator diodes has been joined by several milliamps for computer memories in engine controls, trip computers and entertainment systems, together with several more milliamps for vehicle security systems. Power demands for some of the latter have raised the battery current in nominally 'switched-off' vehicles to over 25 mA in several modern cars. In some cases, this is augmented alarmingly by charging of portable-telephone batteries, which can still occur with the engine switched off.

Whilst these demands are not excessive in a vehicle used on a daily or even weekly basis, in vehicles left in store or in long-term parking at airports they can cause batteries to be discharged to a point where starting is difficult. Moreover, unless some voltage-limiting cut-out is used, it is possible for the battery to become completely discharged in a very damaging way which will involve the majority of the active material on the plates. (Discharges at higher currents are limited to some extent by resistive effects whereas at low currents, exhaustion of the active material is the terminating factor.)

This problem is being approached by limiting the quiescent drain loads and in some cases switching off all but essential loads when the voltage drops below a critical value. It is also possible to economise on some security systems by pulsed rather than continuous operation. There is, however, a need to develop battery structures that will resist the effects of such deep discharges, and dual battery systems mentioned earlier may alleviate the worst risks of battery failure.

A further cure for quiescent drain problems would be to provide some power generator which can be left running safely in unattended vehicles. The only device which comes near to satisfying this requirement at the moment is the silicon solar (photovoltaic) cell. A battery of such cells about 0.5 m^2 in area mounted on the vehicle roof can give useful power input to the system in most climates (Garner, 7.6). Silicon solar cells are, however, still relatively expensive and their input in many parking situations would be negligible. At the time of writing, a small number of add-on systems are being marketed, some being specially devised to power supplementary car ventilation in hot and sunny climates.

7.2 The Future of the Lead-Acid Battery

As mentioned earlier, the lead-acid battery is by no means obsolescent. Improved vehicle electrical architectures will probably maintain its use well into the next century. It will, however, have to change to meet the challenge implied in Section 7.1 and its structure and form may well alter considerably from those in use today. The priorities for improvement may be listed as :

> Increased reliability
> Improved life
> Better charge acceptance
> 'Fit and Forget' - no maintenance versions
> Decreased weight and volume
> Complete recyclability of materials (especially lead)
> Lower cost

Some of these will probably be met by changes in battery structure and some will be met by more intelligent use of batteries and improvements in their operating system.

Some promising changes in battery structure are being examined by the Advanced Lead-Acid Battery Consortium (usually shortened to ALABC). This is an international collaborative research and development body consisting of battery manufacturers, materials suppliers, industrial research corporations and related academic institutions co-ordinated by the International Lead Zinc Research Organisation, with the object of furthering lead-acid battery technology (Cooper, 7.7). Their research programme includes :

- Improvement of positive material utilisation by the use of conducting additions.
- Investigation of electro-osmotic pumping to improve acid access to active material.
- The influence of active material production variables on cycling behaviour at high discharge rates.

- The optimisation of separator materials for VRLA batteries.
- The effects of alloying elements on positive grid life and performance.
- Development of battery material recycling technologies.

Though the main thrust of this work is aimed at traction batteries for electric vehicles, many of the improvements sought will also make a significant impact on automotive SLI batteries.

The contributions of potential structural changes will now be reviewed in more detail and it is convenient to examine the possibilities for individual cell components.

7.2.1 Active Materials

Poor utilisation of the active materials has been a characteristic of conventional lead-acid design. It is noted in Chapter 2 that this in some ways allows the volume changes and mechanical stresses to be accommodated, but it is important to see what improvements can be made before mechanical effects intervene. Moseley (7.8) notes the improvements which might be had by using solid foil substrates instead of the usual grids. Bullock (7.9) notes that utilisation in conventional structures is limited by acid diffusion to the reaction site (impeded by increasing sulphate precipitation) and that improved diffusion or forced flow improves the utilisation. Mechanically assisted flow is not practical in automotive batteries (though it has been suggested in lead-acid batteries for electric vehicles (Borger and Hullmeine, 7.10)) but acid access can be improved by greater porosity and pore sizes. These factors would also increase the fragility of the plate and decrease the resistance to stresses in charge and discharge unless additional support were available. Such support can sometimes be provided by restraining the whole cell assembly under pressure or by the use of tubes or pockets which keep the active material in contact with the support grid (as in 'tubular' traction batteries).

The electro-osmotic pumping referred to previously (Cooper 7.7) uses additions of graphite (McGregor, 7.11) or sulphonated polyvinylidene di-fluoride (SVPDF). Such materials enhance the electro-osmotic pumping effect which exists when an electrical potential is applied across a microporous layer containing ionic solutions. Whilst significant improvement is measured in simple tests, its effectiveness over the life of batteries in the field has yet to be determined.

From time to time, several attempts have been made to add various structural additives to the active materials (McGregor, 7.11) in order to increase conductivity (e.g. by the addition of carbon fibres or graphite) or by polymer-bonding materials to increase mechanical strength. These efforts are usually frustrated by the corrosive environment and the highly oxidising conditions found within the positive plate (where extra conductivity might be particularly appreciated). More recently the use of conducting a-stoichiometric titanium oxides (Cooper, 7.7) has held promise of a stable conducting phase compatible with the working conditions.

The method of preparing active materials (i.e. electro-reduction and oxidation of sulphate pastes) has changed little within this century, though it is possible to produce active porous masses by decomposition of lead compounds in reducing or oxidising conditions, and by pressing chemically-bonded materials with plastic binders. Unfortunately many alternative preparation routes involve substantially different paths for positive and negative active materials. Such changes suffer an immediate economic disadvantage compared to existing methods where all processes but the initial paste mixing are common to both positive and negative plates.

In spite of the difficulties recited above, it is likely that continued efforts will be made to improve active material utilisation for both economic reasons and for the improvement of weight and volume performance figures.

7.2.2 Support Grids

As recorded in Chapter 2, the grids play a vital role in performance and durability. There has been a steady improvement over several decades (Bagshaw, 7.12) in mechanical and electrical design and also in chemical stability. The reduction or elimination of antimony content and the use of calcium alloys were key issues in reducing gas evolution and allowing the development of low-maintenance batteries.

In some respects, pure lead has been seen as ideal for chemical reasons, and it has been used successfully in the Gates battery structure (McClelland et al, 7.13) where its mechanical deficiencies can be accommodated by rolling the plates into a tight 'Swiss roll' structure (see Chapter 1, Figure 1.9). Attempts have also been made in the past to use dispersion-strengthened lead (a material made by rolling together pure lead powder with a small amount of lead oxide) which has the chemical properties of pure lead and considerable mechanical strength and stiffness (Lund et al, 7.14). However, economic production and use of this material pose substantial problems (Bird et al, 7.15), since grid fabrication and recycling the scrap are not as simple as with lead alloys.

More recently, a variety of metal composites have been developed (e.g. for the plates of the Horizon battery, as described in the next section) which involve inorganic strengthening fibres included in lead wires and rods. Again the departure from simple casting processes or continuous strip production means that substantial improvements are required to make such new materials completely acceptable.

The use of electrical simulation tools, capable of analysing voltage-drops over complex resistive networks, has brought a great deal of improvement in grid patterns and terminal connections. A variety of departures from the simple square or rectangular pattern and corner connecting-lug can be found in modern automotive batteries. In the case of the negative electrode, where the active material has substantial electrical conductivity on its own account, some manufacturers have experimented with plastic reinforcement of a lighter lead alloy structure (Rusin, 7.16).

7.2.3 Cell Configurations

The possibility of more specialised battery functions in multiple battery systems, as described in Section 7.1, suggests that new designs of battery may be directed to either the starter battery or service load battery application. Starter batteries will require high power from low capacity, and service load batteries will require good energy storage densities and resistance to deep cycling effects.

The starter battery may utilise the high current design produced by Dunlop (Section 7.1), but the ideal configuration would be a bipolar battery as illustrated in

Chapter 2, Figure 2.6(c). This allows minimum resistive losses in intercell current paths and a shorter current path to the active material. The problem in the lead-acid system is the corrosive, oxidising condition at the positive plate. The sheet of material on which the active materials are supported must have virtually metallic conductivity, yet it must be sufficiently inert to corrosive attack to last the battery service-life. It should also be lightweight and strong in the form of flat sheets and it must maintain a good conductive surface in contact with the active material. As noted earlier, there are few materials that fulfil these requirements, and battery researchers are earnestly testing a wide variety of materials in this role.

Lead itself in the form of thin sheets or foils might be considered the ideal material. However lead, lead-antimony and lead-calcium were tried in thin sheet form by Giner and co-workers (7.17). They found that sheet of sufficient thickness to provide good corrosion resistance gave an excessively heavy structure. An alternative structure tried was polypropylene sheet with lead strips wrapped around so as to provide continuous contact with both sides. Rowlette (7.18) proposed plastic sheet with lead pellet inserts. Rippel (7.19) used polythene with embedded graphite fibres and with the outer surfaces lead-coated.

The search for a lightweight, stable and conductive electrode support sheet is now the prime target of many workers. Even when this is found, the production of thin bipolar cells will pose several challenges in material sealing and joining processes. The gains to be realised can be seen in Figure 7.3, which illustrates the performance of a battery developed by Johnson Controls (Vidas et al, 7.20). In the meantime, other structures with high power-capabilities have been examined.

One novel design was found in the Cathanode™ battery by GNB. This used large numbers of small plates placed at right angles to the usual orientation (see Figure 7.4). Each cell had 66 small plates as compared to less than 20 large plates found in conventional designs. The small-plate configuration together with improved inter-cell connections gave 40% increases in cold-cranking ability as compared to similar batteries of normal construction. The other commercialised high-rate design is the Dunlop Pulsar™ (Howlett, 7.4) shown in Figure 7.1. Each plate is connected through the cell wall to the complementary plate in the adjoining cell, thereby providing a configuration with low internal resistance.

The 'Horizon Battery' (Kline, 7.21) is a novel design using plates disposed horizontally, giving reduced acid stratification and better active-material retention. The grids are based on wires formed by lead-tin alloy coated glass fibre and, like the Dunlop design, individual connections are provided between plates in adjacent cells. The energy capacity achieved is high (50 Wh/kg in 0.33C discharge at 90°F) and discharge power is over 200 W/kg (30 secs).

A very high-power design is provided by the Bolder battery (7.22) shown in Figure 7.5. This is a Swiss roll structure but the ends of the structure provide a massive connection with outer edges of one of the rolled plates. The active materials are in thin layers spread over the surface of the foil supports. This provides an extremely low-resistance configuration at the expense of a lower ratio of active material to support material. The result is a high continuous power density of 800 W/kg and reduced energy density (~30 Wh/kg). Powers of 5000 W/kg can be achieved for a few seconds.

It is apparent that by the use of such novel plate and cell configurations both the performance and durability of the lead-acid battery can be improved still further. Parallel improvements in related electrical components together with the availability of new materials and surface treatments resistant to cell working-conditions are expected to help in the process. If these are combined with more intelligent charging systems and discharge management, there is no reason why lead-acid should not have a promising future in the automotive application. Its low cost materials and production processes give it a head-start on the competition.

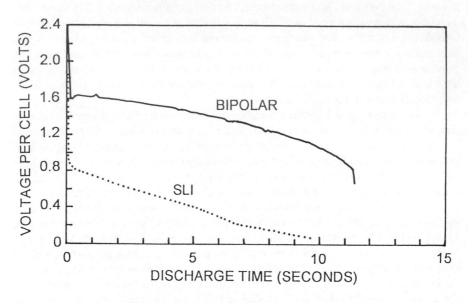

Fig.7.3. Johnson Controls bipolar discharge plot

7.3 Alternative Battery Systems

Since the 1960s, a large amount of effort has been expended on research and development in both primary and secondary battery systems. One reason for this has been the increasing interest in battery-powered road-vehicles, brought about by the desire to exploit renewable energy sources and by an increasing awareness of the pollution effects due to existing road-vehicles. A whole new spectrum of secondary battery chemistries has been examined and a relatively small number have found their way to more general usage.

At the other end of the power and energy scale, the micro-miniaturisation of many electronic devices has brought a need for small, compact, and cheap power sources. This has brought batteries using lithium and more specialised systems such as zinc-air to the point of general commercial acceptance. Batteries used in computers, cameras and similar devices have also brought about renewed interest in battery state-of-charge measurement and in more precise management of re-charge. At the time of writing there are moves to produce the 'smart battery' which has its own integrated microprocessor to protect it from over-discharge and incorrect charging whilst recording

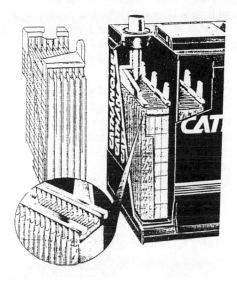

Fig. 7.4. Gould Cathanode battery design
(Gould National Batteries Inc.)

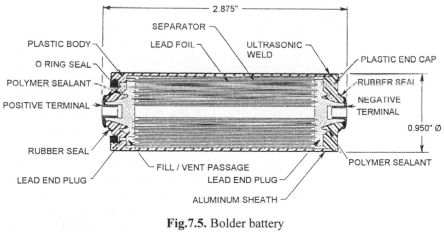

Fig.7.5. Bolder battery
(Elsevier Science SA)

the battery history and capacity deterioration via an algorithm mimicking battery behaviour. The processor would exchange information with the device or system being powered, to produce a complete power-management system.

7.3.1 New Secondary Battery Chemical Systems

In theory, the most energetic battery systems would use materials such as lithium, sodium or potassium with very negative reduction potentials as negative electrodes, and materials such as fluorine or chlorine with high positive reduction potentials as the positive materials (Jasinski, 7.23). Quite apart from the gaseous nature of the latter, such active materials present difficulties in that they attack most common electrolytes (the similar problems with lead and lead dioxide were reviewed in Chapter 2). High-energy active materials therefore demand special electrolytes stable at both very high and very low electrode potentials. Some organic solvents are sufficiently stable but the resulting electrolyte solutions have considerably lower conductivity than the common electrolytes based on water. Molten salt mixtures are good conductors and generally more stable but the temperatures involved are obviously disadvantageous. Some batteries also use conducting ceramics and polymers which also have good stability. The conductivities of various electrolyte options are shown in Table 7.1.

Table 7.1. Electrolyte conductivities

Electrolyte Materials	Conductivity Range $(\Omega^{-1}.cm^{-1})$
Aqueous acid/alkali	10^{-2} to 1
Organic solvents	10^{-4} to 10^{-2}
Molten salts	10^{-1} to 10^{2}
Solid ion conductors (ambient temperature)	10^{-8} to 10^{-4}

The ubiquitous use of automotive batteries precludes the use of high temperatures required by molten salts, and the only practical options appear to be aqueous electrolytes or the more conducting organic solvents solutions such as those based on organic carbonates, nitriles or esters, with dissolved salts such as alkali hexafluorophosphates, or fluoroborates. (Perchlorates are also sufficiently conductive but result in explosive mixtures!)

It can be seen that aqueous electrolytes are still likely to produce the highest power systems on the basis of sheer conductivity. The high-energy silver oxide/zinc and silver oxide/cadmium batteries have occasionally been used for military vehicles and racing cars but cost and material supply problems prohibit their more general use. Nickel-alkaline developments have provided interesting candidates in nickel-zinc and nickel-hydride batteries. The nickel positive electrode forms a good basis for a variety of high-power systems.

A longer-term candidate for automotive purposes is the so-called lithium ion battery. As noted earlier, lithium is a potentially high-energy material but is only stable in organic electrolytes. It suffers from other problems, notably the formation of quasi-passive films on its surface and its re-plating from solution in dendritic form on re-charge. These problems have dogged attempts at lithium secondary batteries over many years (Gabano, 7.24). In the meantime, small primary lithium batteries have made practical the sub-miniaturisation of a whole range of electrical appliances from watches to small computers. They are already used in automotive systems for the long-term support of semiconductor memories when the main power supply is turned off. The load current is vanishingly small (~ 1 μA at 3 V) and cells of up to 1 Ah capacity are built into the subsystems involved, their life expectancy matching that of the vehicle. The lithium ion battery avoids the use of metallic lithium with its attendant problems by the use of lithium intercalates, i.e. compounds absorbing lithium atoms within specialised crystallographic structures.

The alkaline battery candidates and the lithium ion batteries will now be described in more detail.

7.3.2 Nickel-Alkaline Batteries

As noted in Chapter 6, nickel-cadmium batteries have been used in the past. Several related systems, notably nickel-iron, nickel-zinc and nickel-hydrogen (Warthmann, 7.25), have been investigated for electric vehicle use. Nickel-zinc has been used for some time in highly specialised applications. It has high power capability but its cycle life is limited by the behaviour of the zinc electrode which is normally produced in a porous mass by forming a paste of zinc oxide on a current collector sheet or mesh. In alkaline electrolytes, zinc compounds are highly soluble; and on re-charge, zinc electrodes tend to form dendritic growths from the surface which can short-circuit the electrodes. The porosity of zinc electrodes also diminishes on cycling, with a subsequent loss of capacity. These problems can be remedied to some extent by additions of graphite and metal oxides to the negative active material, and by the use of separators which obstruct the growth of dendrites (Duffield et al, 7.26). However, nickel-zinc batteries have significantly shorter cycle-lives than other battery systems. The poor cycling behaviour is exacerbated by float-charging as found in vehicle systems. Nickel-zinc does, however, have the lowest-cost materials of the nickel-alkaline group and if the zinc life problems can be overcome, it could find a place in automotive application.

In the last decade, a variety of nickel-hydrogen battery known as the nickel-hydride system has been developed to the point of commercial availability and is now widely used in low-capacity versions. Large batteries are beginning to find their way into vehicle traction systems. In this battery, the cadmium electrode of nickel cadmium is replaced by an alloy capable of absorbing large amounts of hydrogen at relatively low pressure. The hydrogen is consumed in discharge and regenerated on the electrode surface in re-charge whilst the alloy remains essentially unchanged. The development of the battery as a replacement for nickel-cadmium in many applications has been spurred on by environmental objections to cadmium. The hydrogen storage alloy is a nickel alloy with proprietary additions of rare-earth metals such as lanthanum. For example, LaNi$_5$ stores hydrogen up to the equivalent of 320 Ah/kg at room temperature and 2.5 bar pressure. Companies such as Ovonics (Fetcenko et al, 7.27) have developed proprietary alloys which store large quantities at lower pressure. Unfortunately material costs are high and must be regarded as an inhibitor to large-scale automotive use.

The capabilities of nickel-zinc and nickel-metal hydride batteries are compared with nickel-cadmium in Table 7.2.

Discharge and re-charge voltage profiles are shown in Figures 7.6 and 7.7. All of the alkaline systems above can be made in fully-sealed forms by using oxygen recombination on the negative to avoid internal pressure build-up. Nickel-alkaline batteries tend to suffer higher self-discharge rates than most modern lead-acid types and nickel/metal hydride is the worst in this respect (losing 20% of capacity/month). Self-discharge sets a limit on starting reliability after long inactive parking-spells at airports, etc.

7.3.3 Lithium Ion Batteries
The lithium ion battery derives its name from the fact that lithium ions are exchanged between the positive and negative in charge and discharge (an earlier name was the 'rocking chair battery', since lithium concentration rocked backwards and forwards between the two electrodes). Negative electrodes are made of polymer-bonded graphite

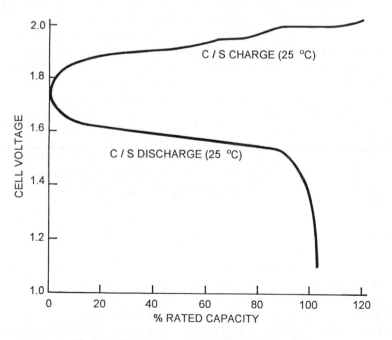

Fig.7.6. Nickel-zinc discharge/charge voltages

Table 7.2. System comparison

System	Wh/kg Theoretical	Wh/kg Realised	W/kg for 30s	Theoretical Cell Volts	Cycle Life
Ni/Cd	209	45	500	1.29	1000
Ni/Zn	324	65	500	1.72	500
Ni/MH	213	53	300	1.31	1000

attached to metal foil or sheet current collectors. Passive films are formed at the surface of the negative which prevent reaction with the electrolyte (self-discharge is low) yet allow the relatively free passage of lithium ions during current flow. Positive electrodes are made from metal oxides (nickel, manganese and vanadium oxides have been used). The type devised by SAFT (Broussely 7.28) for electric- vehicle traction uses the nickel oxide NiO_2. The cell reaction is written as:

$$CLi_x + Li_yNiO_2 \rightarrow CLi_{x-n} + Li_{y+n}NiO_2 \text{ (discharge direction)}$$

It can be seen that during charge the lithium is transferred via lithium ions in solution to the positive (oxide) electrode. The negative electrode may have a composition near LiC_6. The free-energy values of both electrode materials vary with lithium concentration and the cell voltage varies with the state of charge as shown in Figure 7.8.

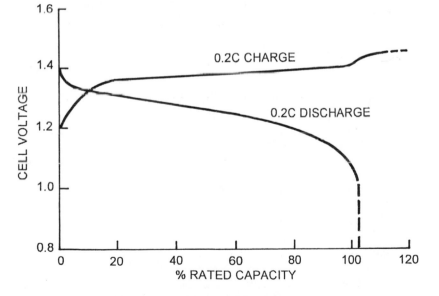

Fig.7.7. Nickel-hydride charge/discharge voltages

The electrolyte used is based on a solution of lithium hexafluorophosphate in propylene carbonate with a variety of other solvents added to improve conductivity and electrode stability. The electrodes are the normal cell configuration of interleaved laminar foils and are separated by porous polypropylene sheets. Containers are of metal and completely sealed.

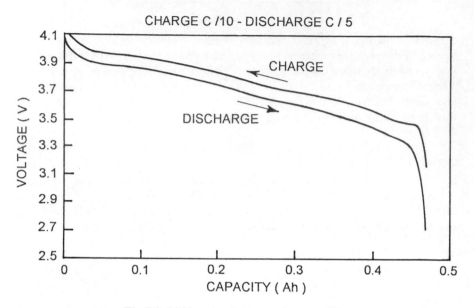

Fig.7.8. Lithium ion battery voltage profile

The operating characteristics given by SAFT are listed below (for a 3 kWh module).

Specific Energy	150	Wh/kg
	250	Wh/litre
Peak Power	300	W/kg
Cycle Life	800	
Target Cost	150	ECU/kWh

To avoid destructive abuse, the voltage limits on cycling must be limited to 2.0 V on discharge and 4.15 V on recharge. Over-discharge results in irreversible change to the nickel oxide positive and overcharge causes current collector corrosion and electrolyte decomposition. Due to the lower electrolyte conductivity, power performance falls off significantly at low temperature.

7.4 Supercapacitors

'Supercapacitors' or 'Ultracapacitors' are special electrochemical cells having electrodes with the highest possible surface area. The electrodes are operated with voltage limits such that either no net reaction or only limited surface reactions take place. Electrodes in contact with an electrolyte form a 'double layer' of charge at the electrolyte/conductor interface, due the ionic charges in solution and the reflected charge distribution at the metal surface. The charge concentration increases with potential, thereby exhibiting some of the characteristics of a classical capacitor, which in normal aqueous solutions amounts to 10-20 µF per square centimetre of electrode surface in contact with the solution (Delahay, 7.29).

In some electrode/solution combinations, potential-dependent specific adsorption of ions can take place on the electrode surface. This considerably increases the effective capacity. In addition, surface reactions involving the deposition of monatomic layers of metals can take place on the electrode surface, and the charge transfer varies with voltage in the same way as a capacitor. Some metal oxides (e.g. ruthenium and iridium oxides) also support redox reactions involving hydrogen which are reversible and which also provide pseudo-capacitor characteristics (Trasatti and Kurzweil, 7.30).

These effects can by used singly or in combination to provide electrolytic capacitors with capacity values a million times higher than similar packages of conventional electrolytic capacitors which rely on the dielectric properties of thin electrolytic oxide films. The resulting devices are known as supercapacitors or ultracapacitors, without any distinction in terms of capabilities, but the former term will be used here.

Supercapacitors differ from batteries in that charge is inserted or withdrawn at a voltage dependent on the amount of charge remaining. The ideal capacitor shows the relationship:

$$V = Q/C$$

where V is the voltage across the capacitor,
Q is the charge stored in coulombs, and
C is the capacity in farads.

In theory, battery charge and discharge occur at a nearly fixed potential which is determined by the electrode reactions. Essentially the same potential is maintained until the active materials are exhausted. (In practice there is a variation of voltage with current due to kinetic 'polarisation' effects but the characteristic potential is usually restored once current flow ceases.) The voltage of a supercapacitor is dependent only on the amount of charge stored (and thereby acts as an indicator of stored energy). This variation of charge or discharge voltage is found with all the electrode processes

180

mentioned above. The discharge characteristics of a battery and supercapacitor are compared in Figure 7.9.

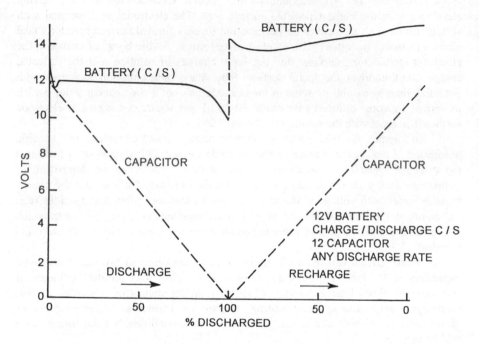

Fig.7.9. Discharge of battery and supercapacitor

The capacitor cells must be kept below voltages at which any undesirable electrode reactions (e.g. electrolyte decomposition) take place and this normally limits them to 1.2 V/cell with aqueous electrolytes or about 3 V/cell with organic electrolytes. Cells may be used in series to produce higher voltage arrays (but the effective capacity reduces in proportion to the number of similar cells in series).

The ionic concentration changes and surface reactions involved in supercapacitors produce no bulk-volume changes as in battery electrodes, and the charge-discharge cycle life is therefore much longer. In addition, if the voltage limits are observed, there should be virtually no corrosion reactions to damage the electrode structures and long calendar lives would be expected. Capacitor cells are normally sealed (with a positive pressure release to accommodate any abuse), and maintenance-free.

The double-layer and absorption processes are subject to some kinetic limitations (c.f. the polarisation effects of battery electrodes). However, the effects produced are relatively small, and the main contribution to voltage changes with current flow comes from resistive effects in the electrodes and electrolyte.

In practical supercapacitor cells, charge is lost through leakage effects, possibly due to unspecified charge-transfer reactions taking place. The purity and stability of electrode and electrolyte materials is therefore critical. Self-discharge effects determine the maximum effective storage-time for the cell system (practical examples show leakage currents of about 20-200 μA/F at full working voltage).

In order to achieve a high capacity, the electrode materials require a high surface area in contact with the electrolyte. A list of candidate materials is shown in Table 7.3.

Table 7.3. Candidate materials for high capacity systems

Material	Specific Surface Area m^2/g	Comments
Raney metals	50-200	Subject to corrosion in some electrolytes
Rare earth oxides	300-500	Chemical stability problems
Carbon blacks	500-1500+	Significant electrical resistance

Not all of the surface indicated for the above materials will be in contact with the electrolyte, since the materials tend to be of a microporous nature and the areas indicated above are measured by gas adsorption. However, very high specific capacities of the order of farads/g can be realised in practice.

The high surface-area materials are spread in thin layers on conducting supports such as thin foils or meshes and may be intermixed with conductivity improvers such as graphite or carbon fibres. Where the devices are specifically designed for high power transfer, it may be necessary to sacrifice energy capacity for improved power by using heavier current collectors and multiple terminal arrangements.

Both electrodes may be nominally the same material and the devices are then potentially polarity insensitive, but most available samples are restricted to a specific polarity.

The devices are not seriously affected by temperature, although electrolyte resistivity increases with lower temperatures. Significantly high temperatures (above 60°C) will reduce electrolyte stability, increase self-discharge rate and probably shorten life expectancy.

The cell designs used for supercapacitors may be the same as those used for batteries (see Chapter 2). Most of the practical examples use the interleaved plate or Swiss roll designs (the latter being standard for conventional electrolytic capacitors). Advanced designs using bipolar configurations are being developed in the USA. Cells are normally sealed from the atmosphere with a positive pressure safety vent.

Dowgallio (7.31) has listed the achievements of developers as shown in Table

7.4. Supercapacitors developed in Russia at the Kursk Accumulator Plant have already been used to start road vehicles, mainly in support of lead-acid batteries (RHISTC, 7.32).

Table 7.4. New developments

Developer	Electrode Material	Wh/kg	Wh/l	Peak W/kg	Cost
Panasonic	Carbon	2.2	2.9	400	Low
Pinnacle Research	Ta,Ru oxides	0.8	3	2000	High
Pinnacle Research	New metal oxides	2-4	10	400	Medium
Maxwell Labs	Carbon-aqueous	1.2	2	800	Medium
Maxwell Labs	Carbon-organic	8	9	2000	Medium
Livermore Nat Lab	Carbon Aerogel	1	1.5	2000	Medium
Sandia Nat Lab	Synthetic Carbon	1.4	1.7	1000	Medium
Los Alamos Nat Lab	Conducting Polymer	10*	?	500*	Low*
Federal fabrics	Carbon	20*	?	500*	Low*

* Projected figures

Longer-term projections of performance, based on materials and developers' claims, are >15 Wh/kg, >2000 W/kg with a discharge cycle-life of over 100,000 at a cost of about $0.5/Wh (Dowgallio, 7.31). If the cost figures can be achieved, then supercapacitors may have a promising future in support of batteries for engine starting. They could be charged steadily from a battery incapable of providing starting power itself (e.g. a lead-acid battery in low state-of-charge) and provide the necessary starting power surge. Their use may also allow batteries to be optimised for other desirable attributes, such as deep cycle life and lower-rate capacity, without sacrificing overall starting capability.

7.5 Which Systems for the Future?

Future vehicle electrical systems may be of a much wider variety than is common at the time of writing (1996). Vehicles using internal combustion engines will probably continue to use one or two lead-acid batteries in the immediate future. Dedicated starter batteries are likely to become more common, particularly as vehicle systems become more complex, and simple alternative means of starting a vehicle become more difficult to apply.

The medium-term future may see changes in engine technology which reduce the starting power necessary. Even existing vehicles with petrol engines and fuel-injection require substantially less starter effort than early types with carburettors. With fully variable valve-timing, it would be possible to avoid compression until the engine was cranking steadily and starter-power could be substantially decreased. The system requirements might then be met with a large deep-cycle battery of a conventional type and a supercapacitor battery for cranking. The deep-cycling battery might well be our old friend lead-acid, in a suitably modified form having well-supported active materials. A lithium ion battery would, however, occupy less space and weight and its state-of-charge could be monitored by voltage measurements.

Alternatively, a small bipolar battery could be used for starting and both lead-acid and alkaline types might be considered in this dedicated application, though materials costs would make the latter less attractive.

In the more distant future, electric-traction systems in zero-emission vehicles may use fuel-cells (which convert gas or liquid fuels directly to electricity) or large secondary batteries as the prime energy supply. Traction supply voltages are likely to be over 200 V for efficient power controller dèsign, and a high efficiency DC/DC converter would be required to provide lighting and auxiliary system supplies. Some kind of lower-voltage auxiliary system battery will therefore be needed to back up controller microprocessor systems, and to be the means of system-activation after long periods of inactivity. The best cars of today are expected to start promptly after 6 months' lay-up.

It was noted in Chapter 1 that the car starter battery has been responsible for keeping secondary batteries in the public eye. It seems likely that the personal need for private transport, and its realisation in the motor car will remain the layman's prime contact with the reality of secondary batteries, and particularly with their failings. Some of the developments described above may well be able to make the motorist have a new image of automotive batteries, or perhaps, even better, to forget their presence altogether. A truly 'fit and forget' item, which performs reliably in all circumstances, and which has a life comparable to that expected of the vehicle, is the dream of the vehicle manufacturer, the electrical-system engineer and the final customer.

REFERENCES

7.1 'Dual Voltage Systems for Future Autos' by B A Jarrett and J G W West, Design News, Oct 1986, p140-146.

7.2 'Vehicle Electrical Power Supply Systems and Their Impact on System Design' by G A Williams and M J Holt, Proc. Inst. Mechanical Engineers, 206(1992), p149.

7.3 'Getting the Best Out of 12 Volts – The Development of an Advanced Electrical Architecture Vehicle' by A Williams, J Smith, H Evans, J Scholfield and J Elvidge, Society of Automotive Engineers Paper SAE 940386, February 1994.

7.4 'High Power Lead-Acid Batteries' by J Howlett, J Power Sources, 11(1984), p43-45.

7.5 'Optimised Batteries for Cars with Dual Electrical Architecture' by J Douady, C Pascon, A Dugastand, G Fossati, J Power Sources 53(1994), p367-375.

7.6 'Car Auxiliary Power Supply by Solar Cells' by I F Garner, Proceedings of AUTOTECH 91, 1991 (Inst. Mechanical Engineers).

7.7 'Collaboration in Research - The ALABC;Brite EURAM Lead-Acid Electric Vehicle Project' by A

184

Cooper, J Power Sources, $\underline{59}$(1996), p161-170.

7.8 'Lead-Acid Battery Myths' by P T Moseley, J Power Sources, $\underline{59}$(1996), p81-86.

7.9 'The Advancing Lead-Acid Battery' K R Bullock in 'Advances in Lead-Acid Batteries', Electrochem. Soc. Proceedings 84-14, 1984.

7.10 'Influence of Lampblack on Capacity Retention in $Pb/PbSO_4$ Electrodes with forced Flow of Electrolyte' by W Borger and U Hullmeine, in 'Advances in Lead-Acid Batteries', Electrochem. Soc. Proceedings 84-14, 1984.

7.11 'Active Material Additives for High Rate Lead-Acid Batteries - Have There Been Any Positive Advances?' by K McGregor, J Power Sources, $\underline{59}$(1996), 31-43.

7.12 'Lead Alloys, Past, Present and Future' by N E Bagshaw, J Power Sources, $\underline{53}$(1995), 25-30.

7.13 McClelland et al. (Gates Energy Products), US Patents 3,704,173 and 3,862,861.

7.14 'Structure and Properties of Dispersion Hardened Lead' by J A Lund, E G von Tiesenhausen, and D Tromans, in 'Lead '65', Proc. 2nd International Conf. on Lead, Pergamon Press, 1967.

7.15 'Construction and Service Testing of Cells made with Grids of Dispersion-Hardened Lead' by T L Bird, I Dugdale, G G Graver, Proc. 3rd Conf. Lead (Lead '68), Venice, 1968, Sponsored by the Lead Development Association (Pergamon Press 1968).

7.16 'New Materials for Current Carrying Components in Lead-Acid Batteries' by A I Rusin, J Power Sources, $\underline{36}$(1991), 473-478.

7.17 'Lead-Acid Battery Development for Heat Engine/Electric Hybrid Vehicles' by J Giner, A H Taylor, F Goebel, USEPA Contract EHSH71-009, Final report, Nov.1971 (APTD 1346).

7.18 J Rowlette (California Inst. Technology) US Patent 4,542,082, 1985.

7.19 W Rippel (California Inst. Technology) US Patent 4,275,130, 1979.

7.20 'The Bipolar Lead-Acid Battery at Johnson Controls Inc.' by R A Vidas, R C Miles, P D Korinek, Proc 34th International Power Sources Symposium, 1990 (IEEE, Piscataway, NJ, USA, 1990).

7.21 'Getting Horizon Ready for the Market', C R Kline, Batteries International, January 1994, p40-41.

7.22 'A New High Rate, Fast Charge Battery' by T Juergens and R F Nelson, J Power Sources, $\underline{53}$(1995), p201-205.

7.23 'High Energy Batteries' by R Jasinski, Plenum Press, New York, 1967.

7.24 'Lithium Batteries' J P Gabano (Ed), Academic Press, London, 1983

7.25 'Sealed Alkaline Batteries for Electric Cars' by W Warthmann, Proc. EVS12, Dec.1994, Anaheim,Volume 2, 1994.

7.26 'Evaluation of Additives for a Secondary Zinc Electrode' by A Duffield, P J Mitchell, D W Shield and N Kumar, in 'Power Sources 11' L J Pearce (ed), International Power Sources Symposium Committee, 1986.

7.27 'Hydrogen Storage Materials for Use in Rechargeable Ni-Metal Hydride Batteries' by M A Fetcenko, S Venkatesan, K C Hong, B Reichman, in 'Power Sources 12, 1988, Academic Press, London.

7.28 'Lithium-Carbon Liquid Electrolyte Battery System for Electric Vehicles' by M Broussely, P Flament, C Morin and G Sarre, Proc.EVT '95, Paris, Nov. 1995 (AEVRE, 1995).

7.29 'Double Layer and Electrode Kinetics' by P Delahay, McGraw-Hill, 1964.

7.30 'Electrochemical Capacitors as Versatile Energy Stores' by S Trasatti and P Kurzweil, Platinum Metals Review, 38(1994), 46-56.

7.31 'Perspective on Ultracapacitors for Electric Vehicles' by E J Dowgallio and J E Hardin, IEEE Systems Magazine, 10, p26.

7.32 'Energy Capacitors for Starting Internal Combustion Engines', Publication I-12, Russian House for International Scientific and Technical Co-operation, Moscow 1993.

APPENDIX 1

Glossary of Terms

Activation Overpotential Contribution to the total overpotential due to the charge transfer step at the electrode interface.

Active Mass or Active Material The material forming part of the electrode(s) that actually takes a part in the electrochemical reactions, and normally changes its chemical form as a result of those electrochemical reactions.

Alternator A rotating machine driven by the vehicle engine, generating alternating current.

Ambient Temperature The temperature of the surroundings in which the item is placed.

Ampere-Hour Capacity The quantity of electricity delivered by a cell or battery as measured in ampere-hours (Ah) with certain specified conditions.

Anode The electrode at which the electrochemical reaction involves oxidation. This is the negative electrode while the cell or battery is being discharged, and the positive electrode of the cell or battery during charge.

Anolyte The electrolytic phase in contact with the anode.

Battery Two or more cells connected electrically in series and/or parallel in order to form a unit which provides the required voltage/current combination. For example, a 12 V SLI battery is made up of six 2 V cells in series. However, the term is also often used to indicate a single cell, particularly in the case of a 'dry battery'.

Binder A polymeric material added to the active mass to increase its mechanical strength.

Bipolar Electrode or Plate A plate construction where the positive and negative active materials are on opposite sides of an electronically conducting plate which performs both functions of cell interconnector and cell wall. Also known as a 'duplex' electrode, especially in Leclanché batteries.

Bus, Bus-Bar A rigid metallic connector which connects different elements of a battery; also, the conductor for an electrical system to which a battery terminal is attached.

Button Cell Miniature cylindrical cell having a characteristic disc shape.

Can or Case The external envelope of a cell or battery, or the box containing the cells and connectors.

Capacity, rated The value of the output capability of a battery, expressed in Ah, at a given discharge rate before the voltage falls below a given cut-off value, as indicated by the manufacturer.

Cathode The electrode at which the electrochemical reaction involves reduction. This is the positive electrode while the cell or battery is being discharged, and the negative electrode of the cell or battery during charge.

Catholyte The electrolytic phase in contact with the cathode.

Cell A unitary electrochemical device for storing electrical energy comprising a positive electrode and a negative electrode immersed in an electrolyte.

Cell Reversal Inversion of the polarity of the terminals of a cell in a muticell battery. Cell reversal is usually due to overdischarge, when differences in cell capacities in a series chain allow a low-capacity cell to be overdischarged even though the whole chain may not have reached cut-off voltage.

Charge An electrochemical process by which electrical energy is converted into chemical energy and stored within a cell or battery.

Charge Acceptance The ability of a secondary cell or battery to convert the active material to a dischargeable form. It is measured by the capacity which can be subsequently delivered to a load as a result of the charging process. If the charge acceptance were 100%, then all of the electrical energy input would become available for useful output.

Charge Retention That part of the capacity of a cell or battery available after prolonged storage under specified conditions.

Charge, state of The condition of a cell or battery in terms of the remaining available capacity.

Collector, Current Collector Electronic conductor embedded in the active mass and connected to the bus bar or terminal.

Concentration Overpotential (Overvoltage) Contribution to the total overpotential due to non-uniform concentrations in the electrolyte phase near the electrode surface caused by the passage of current.

Concentration Polarisation Polarisation that is caused by the depletion of ions in the electrolyte in close proximity to the surface of an electrode.

Corrosion Oxidation of a metallic phase starting at the surface and caused by the reaction of the metal with components of the environment. In batteries, corrosion phenomena play an important role, especially in the case of primary aqueous cells and in high temperature systems.

C-rate A method for expressing the rate of charge or discharge of a cell or battery. A cell discharging at a C-rate of r will deliver its nominal rated capacity in $1/r$ h; e.g. if the rated capacity is 2 Ah, a discharge rate of C/1 correponds to a discharge current of 2 A, a rate of C/10 to 0.2 A, etc.

Creep The process by which liquid electrolytes, and in particular alkalies, can escape past rubber-metal or polymer-metal seals, or through minute cracks in a cell case or lid.

Current Density The current per unit active area of the surface of an electrode. It is generally defined in terms of the geometric or projected electrode area and is measured in A m^{-2} or mA cm^{-2}.

Cut-Off Voltage The voltage at which the discharge or charge of the cell or battery is terminated, and beyond which little benefit can be gained. In the case of discharge, it is chosen as the voltage value below which the connected equipment will not operate, or below which operation is not recommended because of the onset of irreversible processes in the cell. In the case of charge, it is selected to allow complete conversion of active material with a minimum of gassing.

Cycle A unit of specified charge and discharge of a cell or battery from and to a set condition.

Cycle Life The total number of charge/discharge cycles that can be delivered by a secondary cell or battery, while maintaining a pre-determined output capacity and cycle energy efficiency.

Depolariser A substance which is supposed to reduce electrode polarisation.

Diffusion The movement of ions in an electrolyte from areas of high concentration to areas of low concentration.

Discharge An electrochemical process by which chemical energy is converted into electrical energy within a cell or battery.

Discharge Curve A plot of cell or battery voltage as a function of time, or of discharge capacity, under a defined discharge current or load.

Discharge Depth The percentage of the capacity to which a cell or battery has been disharged. Shallow/deep discharge = small/large fraction of the usable capacity consumed.

Drain Withdrawal of current from a cell or battery.

Dry Cell A cell containing an immobilised electrolyte, where the electrolyte is either in the form of a paste or gel, or absorbed in a microporous separator material. Normally used to describe cells of the Leclanché type.

Dry-Charged Battery A battery in which the electrodes are in a charged state, but lacking the electrolyte required for the battery to function.

Duplex Electrode A type of electrode consisting of a sheet of conducting material with a layer of positive active material on one side, and a layer of negative active material on the other. (Also bi-polar electrode.)

Dynamo A rotating machine driven by the vehicle engine, generating direct current.

Electrode The part of a cell or battery comprising a current collector and active material at which an electrochemical reaction can take place.

Electrodeposition Deposition of a chemical species at the electrode of an electrolytic cell caused by the passage of electric current.

Electrolysis Chemical modifications (i.e. oxidation or reduction) produced by passing an electric current through an electrolyte.

Electrolyte The medium through which ions can be transported between the positive and negative electrodes of a cell or battery. It may be solid or liquid. In some cases the electrolyte may take part in the cell reaction.

Energy Density The ratio of the energy available from a cell or battery, normally expressed in watt hours, to its volume, normally expressed in litres, or its weight, normally expressed in kilograms.

Equalising Charge Passage of an amount of charge by which the undercharged cells of a battery are brought up to a fully charged condition without damaging those already fully charged.

Expander A substance added in small amount to the active materials of a lead-acid battery to improve the service life and capacity of the electrodes. In particular, an expander prevents the increase in crystal grain size of lead in the negative electrode.

Failure The state in which the performance of a cell or battery does not meet the normal specifications.

Float Charging A method of re-charging in which a secondary battery is continuously connected to a constant voltage supply that maintains the cells in fully charged condition.

Formation or Forming The electrochemical process by which the material on an electrode is converted into active material at the manufacturing stage.

Fuel Cell An electrochemical device in which the active materials are continuously supplied and the reaction products continuously removed, giving the ability to provide an electric current.

Gassing Gas evolution which takes place at the electrodes towards the end of the charging of a cell or battery.

Generator A general term used for the device for generating electricity in a vehicle. It is used to mean either the alternator or the dynamo.

Grain Boundary The surface separating two regions of a solid material which have different crystal orientations or compositions.

Grid The lead framework of a lead-acid battery plate which holds the active material in place and which also acts as the current collector.

Group A set of electrodes within a cell which are connected in parallel.

High Tension The high voltage required to generate the arc across the terminals of the spark plug. Obtained from the secondary winding on the coil.

Hybrid Cell Electrochemical cell in which one of the two active reagents is in the gas phase and may be supplied from an external source. A hybrid cell occupies an intermediate position between closed cells and fuel cells.

Immobilised Electrolyte See Dry Cell.

Inhibitor A substance added to the electrolyte which prevents an electrochemical process, generally by modifying the surface state of an electrode. A well-known example is that of corrosion inhibitors which prevent metal corrosion.

Initial Drain Current that a cell or battery supplies when first placed on a fixed load.

Internal Resistance The resistance to the flow of current within a cell or battery, consisting of both the ionic and electronic resistances of all the components in the electrical path, causing a drop in closed-circuit voltage proportional to the current drain from the cell.

iR loss ; iR drop Decrease in the voltage of a cell during the passage of current, due to the internal resistance of the bulk phases within the cell - mainly that of the electrolyte and the separators. Also known as 'ohmic loss'.

Load The external devices or circuit elements to which electric power is delivered by a cell or battery.

Load Levelling The intervention aimed at reducing non-uniform conditions in electric demand. The principle of load-levelling is to store energy when demand is low and to use it to meet peak demand. Assemblies of lead-acid batteries have been used for this purpose with considerable success.

Local Action Electrochemical reactions that take place at electrodes that dicharge the electrode without any current flowing externally.

Maintenance The procedures which are required in order to keep a battery in proper operating condition. They may include trickle-charging to compensate for self-discharge, addition of water to electolyte, etc.

Maintenance-Free Battery A battery that does not require maintenance, such as additions of water, during its useful life.

Mass Transport Transfer of materials consumed or formed in an electrode process to or from the electrode surface. Mechanisms of mass transport may include diffusion, convection and electromigration.

Migration The movement of ions under the influence of a potential gradient.

Negative Negatively charged electrode, usually of a secondary cell; acts as anode during discharge and cathode during charge.

Ohmic Loss See iR loss.

Open-Circuit Voltage The potential difference between the two terminals of a cell or battery when the external circuit is open, i.e., no current is passing. It may be measured with a high-impedance voltmeter or potentiometer.

Overcharge The application of current from a higher voltage source to a cell or battery which has already reached a fully charged condition.

Overdischarge The forced discharge of a cell or battery beyond 100% of the available capacity. In the case of a multi-cell battery, the overdischarge may cause **cell reversal.**

Overpotential ; Overvoltage Difference between the actual electrode voltage when a current is passing, and the equilibrium (zero current) potential. A number of different effects may contribute to the total overvoltage.

Oxidation The loss of electrons by a chemical species.

Parallel When like terminals of like cells or batteries are connected together to form a battery of larger capacity with the same voltage.

Passivation The creation of a protective layer on the surface of a metal, which, without that layer, would otherwise be reactive.

Passivity The condition of a metallic material corresponding to an immeasurably small rate of corrosion.

Pasted Plate The term used to describe an electrode where the active material, or its precursor, as a paste, has been applied to the grid support member during manufacture.

Polarisation Deviation from equilibrium conditions in an electrode or galvanic cell caused by the passage of current. It is related to the irreversible phenomena at the electrodes (electrode polarisation) or in the electrolytic phase (concentration polarisation).

Polarisation Loss Reduction in the voltage of a cell, delivering current, from its equilibrium value.

Positive Positively charged electrode, usually of a secondary cell ; acts as cathode during the discharge and anode during the charge.

Post See Terminal.

Power Density The ratio of the the power available from a cell or battery, normally expressed as watts,to its volume, in litres, or its weight, in kilograms.

Primary Cell or Battery A cell or battery which cannot, to all intents and purposes, be recharged by the application of an electric current, and is discarded when all its capacity has been used.

Recombining Cell A secondary cell in which provision has been made for the products of overcharge reactions to recombine so that no net change occurs to the composition of the cell system as a result of overcharging.

Rectifier A device for obtaining direct current from a source of alternating current.

Reduction The gain of electrons by a chemical species.

Regulator A device for controlling the voltage of the electrical energy obtained from the generator, as it is supplied to the battery.

Reserve Battery In principle, any battery which will not deliver current in its manufactured form until activated by a suitable procedure, e.g., by adding the electrolyte to the dry components (water-activated cells or cells activated by the addition of special electrolytes), or by raising the temperature of the cell (thermal batteries, where the electrolyte is generally a mixture of salts in the solid state at ambient temperature and liquefy on heating).

Reversal See Cell Reversal.

Secondary Cell or Storage Battery A cell or battery which, after discharge, can be recharged by the passage of an electric current in the opposite direction to that of discharge.

Self-Discharge The loss of capacity due to internal discharge reactions that take place within the cell or battery. See Local Action.

Self-Discharge Rate The rate at which a cell or battery loses service capacity when standing idle.

Separator Electrically insulating layer of material which physically separates electrodes of opposite polarity. Separators must be permeable to the ions of the electrolyte and may also have the function of storing or immobilising the electrolyte.

Series When unlike terminals of like cells or batteries are connected together to form a battery of greater voltage and the same capacity.

Service Life Timescale of satisfactory performance of a battery under a specified operating schedule, expressed in units of time or number of charge/discharge cycles.

Shedding The process whereby poorly adhering active mass (generally in the positive plate of a lead-acid cell) falls from the grid to form a sludge (mud) on the floor of the cell.

Shelf Life The time that a battery can be stored under specified conditions and still meet a specified performance.

Short Circuit The condition when the terminals of a cell or battery are connected directly, allowing a high current to flow. This also covers the situation where, within a single cell, a crystal of lead grows through the separator, or a portion of active material which has been shed from an electrode forms a 'bridge' between plates of opposite polarity round the edge of a separator, and an 'internal short-circuit' is formed.

SLI Battery A battery , usually of 12 V or 24 V, used for engine Starting, Lighting and Ignition in vehicles with internal combustion engines.

Stack An assembly of parallel plates.

Storage Battery See Secondary Battery.

Surface Active Agent A substance which modifies the behaviour of a phase by interacting with its surface. For example, in the case of lead-acid batteries, the morphology of the active materials deposited at the electrodes may be strongly affected by the addition of surface-active agents.

Terminal The external electrical connection of a cell or battery; also known as 'terminal post' or 'post'.

Thermal Battery A type of reserve cell which is activated by raising the temperature.

Thermal Management The means whereby a battery system is maintained within a specified temperature range whilst undergoing charge or discharge.

Thermal Runaway A process whereby a cell or battery on either charge or discharge can destroy itself as the heat generated by the passage of current reduces the internal resistance, thereby enabling an ever-increasing current to be passed.

Trickle Charge A very low rate of charge, sufficient to counter any self-discharge effects and maintain the battery in a fully charged condition.

Uninterruptible Power Supply (UPS) A power system, normally consisting of a bank of batteries with appropriate means for maintaining them fully charged and with an automatic switching arrangement, which maintains current flow without even a momentary break, even in the event of mains or generator failure.

Vent Valve mechanism which allows controlled escape of gases generated during charging, but prevents spillage of electrolyte.

Voltage Delay Time interval at the start of a discharge during which the working voltage of a cell is below its steady value. The phenomenon is generally due to the presence of passivating films on the negative electrode.

Wet Cell A cell in which the electrolyte is free-flowing, and is not wholly absorbed within the separator.

APPENDIX 2

Numerical Data

Some Physical Properties of Lead

Atomic Weight	207.2
Atomic Number	82
Crystal Structure	Face-centred cubic
Valencies	2 & 4
Specific Gravity (20°C)	11.34
Melting Point (°C)	327.4
Boiling Point (°C)	1751
Resistivity (μohm cm, 20°C)	20.65
Tensile Strength (kg/cm)	126.55-175.77
Brinell Hardness (cast)	4.2
Thermal Conductivity (cal $s^{-1}cm^{-2}$,°C/cm, 20°C)	0.083
Magnetic Susceptibility 10^{-6}cgs units	-0.12
Modulus of Elasticity 10^6kg cm^{-2}	0.155
Latent Heat of Vaporisation cal g^{-1}	204
Latent Heat of Fusion cal g^{-1}	5.89
Linear Coefficient of Expansion °C 10^{-6} units	29.3

Faraday's Laws of Electrolysis

1. The product of electrolysis is directly proportional to the amount of electricity passed.
2. Quantities of different substances produced or consumed by the action of a given quantity of electricity are in direct proportion to their equivalent weights.

One faraday = one chemical equivalent weight

One faraday = 96500 coulombs (1 amp for 1 second)

One amp.hour (Ah) = one amp for one hour

26.8 Ah = one faraday

Equivalent weight = atomic weight / valency

Relating this to the lead-acid battery for which the accepted generalised charge/discharge equation is:

$$Pb + PbO_2 + 2H_2SO_4 = 2PbSO_4 + 2H_2O$$

One Ah of discharge converts 3.86 g of lead, 4.46 g of lead dioxide and 3.66 g of sulphuric acid into 11.30 g of lead sulphate and 0.67 g of water. Of course, on charge the reverse happens.

Atomic weight of lead = 207
Molecular weight of lead dioxide = 239
Molecular weight of sulphuric acid = 98
Molecular weight of lead sulphate = 303
Molecular weight of water = 18

Weight of sulphuric acid formed or decomposed during charge (discharge) of a battery
= 3.66 g x no. of Ah charged (discharged)

Weight of water decomposed or formed during charge (discharge) of a battery
= 0.672 g x no. of Ah charged (discharged)

Net change in weight of the electrolyte during charge (discharge)
= 2.987 g x no. of Ah charged (discharged)

Physical Properties of Lead Oxides

Property	PbO	PbO_2	Pb_3O_4
Molecular Weight	223.21	239.19	685.57
Colour	-Red -Yellow	Dark brown or black	Orange to brick red
Structure	-Tetragonal -Orthorhombic	Orthorhombic Tetragonal	Spinel
Density (g/cm^3)	9.2-9.5 9.5-9.9	Reported 9.165, 9.375 and 9.4 for $PbO_{1.919}$	9.1
Melting Point	897°C, sublimes before melting	Decomposes at 290°C	830°C (in oxygen pressure) decomposes in atm. at 500°C
Electrical Properties	p or n type semiconductor	Semiconductor (resistivity at 20°C 91 M cm)	Non-conductor
Solubility	0.5 g l^{-1} in H_2O at 25°C 0.107 g l^{-1} in H_2O at 25°C	Insoluble in cold and hot water. Soluble in HCl. Slightly soluble in HNO_3 and H_2SO_4	Insoluble in cold and hot water. Soluble in HCl and HNO_3 and warm conc. H_2SO_4

Source: Journal of Power Sources, Vol. 2, No. 1, October 1977, Elsevier.

Standard Electrode Potentials

Couple	E^0/V
$Li^+ + e^- = Li$	-3.05
$Ca^{2+} + 2e^- = Ca$	-2.87
$Na^+ + e^- = Na$	-2.71
$Al^{3+} + 3e^- = Al$	-1.66
$2 H_2O + 2e^- = H_2 + 2OH^-$	-0.83
$Zn^{2+} + 2e^- = Zn$	-0.76
$Cr^{3+} + 3e^- = Cr$	-0.74
$PbSO_4 + 2e^- = Pb + SO_4^{2-}$	-0.36
$Ni^{2+} + 2e^- = Ni$	-0.25
$Pb^{2+} + 2e^- = Pb$	-0.13
$Fe^{3+} + 3e^- = Fe$	-0.04
$2H^+ + 2e^- = H$	0
$Cu^{2+} + 2e^- = Cu$	0.34
$O_2 + 2H_2O + 4e^- = 4OH^-$	0.40
$Pt^{2+} + 2e^- = Pt$	1.20
$O_2 + 4H^+ + 4e^- = 2H_2O$	1.23
$Cl_2 + 2e^- = 2Cl$	1.36
$Au^{3+} + 3e^- = Au$	1.40
$Pb^{4+} + 2e^- = Pb^{2+}$	1.67
$F_2 + 2e^- = 2F^-$	2.87

Antelman(1982), The Encyclopedia of Chemical Electrode Potentials, Plenum, New York.

APPENDIX 3

Technical Data

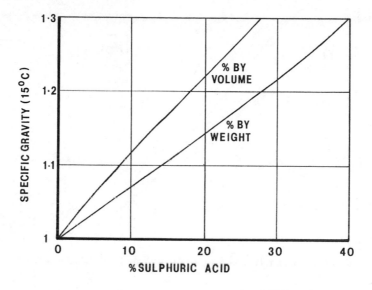

Fig.App3.1. Relationship between specific gravity (15°C) and percentage of sulphuric acid in solution

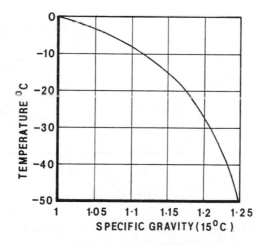

Fig.App3.2. Freezing point of sulphuric acid solutions

202

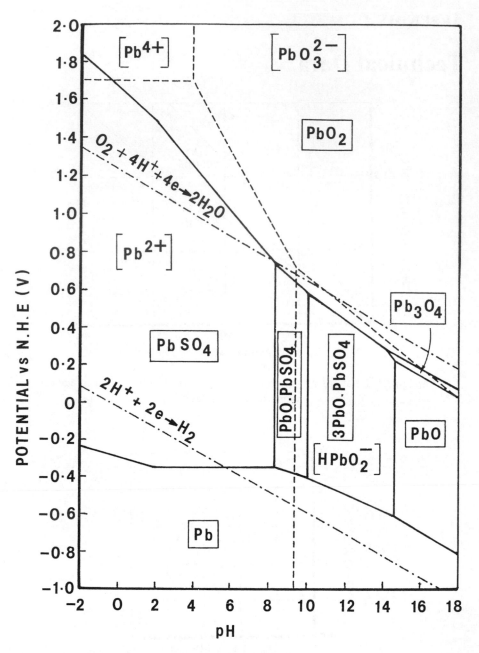

Fig.App3.3. Potential - pH diagram of lead in the presence of sulphate ions at unit activity and at 25°C

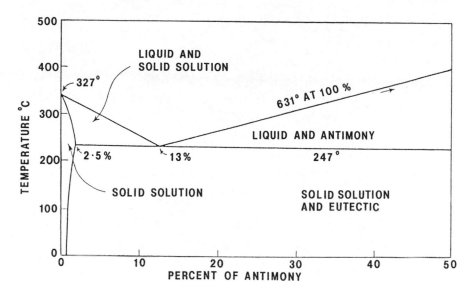

Fig.App3.4. Equilibrium diagram of lead-antimony
(Source: G.W.Vinal, 'Storage Batteries', Wiley, New York, 1955, p.18)

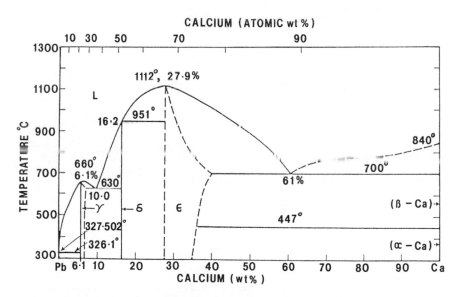

Fig.App3.5. Phase diagram of lead-calcium
(Source: Metals Handbook, 8th edn, Vol. 8, American Society for Metals, Cleveland, Ohio, 1973, p.280)

APPENDIX 4

Bibliography

BOOKS

Adams, R.N., *Electrochemistry at Solid Electrodes,* Dekker, New York, 1969.

Albery, W. J., *Electrode Kinetics,* Clarendon Press, Oxford, 1975.

Bagotzky, V. S. & Skundin, A. M., *Chemical Power Sources*, Academic Press, London, 1980.

Bagshaw, N. E., *Batteries on Ships,* Research Studies Press, Taunton, 1982.

Barak, M. (Ed.), *Electrochemical Power Sources (Primary and Secondary Batteries),* P Peregrinus, Stevenage, 1980. (Recently re-issued by the Institution of Electrical Engineers.)

Berndt, D., *Maintenance-Free Batteries, Lead-Acid, Nickel/Cadmium, Nickel/Metal Hydride, A Handbook of Battery Technology, Second Edition,* Research Studies Press, Taunton, 1997.

Bockris, J. O'M. & Reddy, A. K. N., *Modern Electrochemistry,* Volumes 1 and 2, MacDonald Technical and Scientific, London, 1970.

Bockris, J. O'M., Conway, B. E., Yeager E. & White, R. E., *Electrochemical Energy Conversion and Storage*, Comprehensive Treatise of Electrochemistry, Volume 3, Plenum Press, NewYork, 1981.

Bode, H., *Lead-Acid Batteries,* Wiley, New York, 1977.

Dasoyan, M. A. & Aguf, Z., *Current Theory of Lead-Acid Batteries (English Translation),* Technicopy Ltd, Stonehouse 1980.

206

Falk, S. & Salkind, A. J., *Alkaline Storage Batteries*, Wiley, New York, 1969.

Jasinski, R., *High-Energy Batteries*, Plenum Press, New York, 1969.

Jensen, J., McGeehin, P. & Dell, R., *Electric Batteries for Energy Storage and Conservation*, Odense University Press, Odense, 1979.

Kordesch, K. V., *Batteries: Volume 2, Lead-Acid Batteries and Electric Vehicles*, Dekker, New York, 1977.

Korita, J., Dvorak J. &. Bohackova, V., *Electrochemistry*, Methuen, London, 1970.

Kortum, G., *Treatise on Electrochemistry*, Elsevier, Amsterdam, 1965.

Kuhn, A.T., *The Electrochemistry of Lead*, Academic Press, London, 1979.

Linden, D., *Handbook of Batteries and Fuel Cells*, McGraw-Hill, New York, 1984.

Pourbaix, M., *Atlas d'Équilibres Electro-Chimiques*, Gauthier-Villars, Paris, 1963.

Smith, G., *Storage Batteries*, Isaac Pitman & Sons, London, 1964.

Thirsk, H. R. & Harrison, J. A., *Guide to the Study of Electrode Kinetics*, Academic Press, London, 1972.

Treadwell, A., *The Storage Battery*, Whittaker & Co, London 1898.

Vetter, K. J., *Electrochemical Kinetics*, Academic Press, London,1967.

Vinal, G. W., *Storage Batteries*, McGraw-Hill, New York, 1955.

Witte, E., *Blei-und Stahlakkumulatoren*, Krausskopf-Verlag, Mainz, 1967.

PROCEEDINGS etc.

Bockris, J. O'M. & Conway, B. E., *Modern Aspects of Electrochemistry*, Volumes 1 to 8, Butterworth Scientific, London, 1954-74, Volumes 9 to 15, Plenum Press, New York, 1975-83.

Collins, D. H., Batteries, 1962, Pergamon Press, Oxford 1963.

Collins, D. H., Batteries 2, 1964, Pergamon Press, Oxford, 1965.

Collins, D. H., Power Sources, 1966, Pergamon Press, Oxford, 1967.

Collins, D. H., Power Sources 2, 1968, Pergamon Press, Oxford, 1969.

Collins, D. H., Power Sources 3, Oriel Press, Newcastle-upon-Tyne, 1971.

Collins, D. H., Power Sources 4, 1972, Oriel Press, Newcastle-upon-Tyne, 1973.

Collins, D. H., Power Sources 5, Academic Press, London, 1975.

Collins, D. H., Power Sources 6, Academic Press, London,1977.

Delahay, P. & Tobias, C. W., *Advances in Electrochemistry and Electrochemical Engineering,* Volumes 10 to 13, Wiley, New York, 1977-81.

Gerische, H. & Tobias, C. W., *Advances in Electrochemistry and Electrochemical Engineering,* Volumes 10 to 12, Wiley, New York, 1977-81.

Thompson, J., Power Sources 7, Academic Press, London,1979.

Thompson, J.,Power Sources 8, 1980, Academic Press, London, 1981.

Thompson, J., Power Sources 9, 1982, Academic Press, London, 1983

Thompson, J., Power Sources 10, 1984, Academic Press, London, 1985.

Yeager, E. & Salkind, A. J., *Techniques of Electrochemistry,* Volumes 1 to 3, Wiley, New York, 1973 - 8.

JOURNALS

Advanced Battery Technology, Seven Mountains Scientific Inc., (Boatsberg, P.A.)

Batteries International, Euromoney Publications plc, (Rottingdean).

Battery Man, Battery Council International.

Electrochimica Acta, International Society for Electrochemistry, Pergamon Press, Oxford.

Electrokhimiya (English Translation), Consultants Bureau, New York.

EPRI Journal.

Journal of Applied Electrochemistry, Chapman and Hall, London.

Journal of Electroanalytical Chemistry, Elsevier Sequoia S.A., Lausanne.

Journal of Power Sources, Elsevier Sequoia S.A., Lausanne.

208

Journal of the Electrochemical Society, The Electrochemical Society, Princeton, N. Jersey.

Solid State Ionics, North Holland, Amsterdam.

Index